Get Ready for Microbiology

Lori K. Garrett

DANVILLE AREA COMMUNITY COLLEGE

Judy Meier Penn

SHORELINE COMMUNITY COLLEGE

Benjamin Cummings

San Francisco   Boston   New York   Cape Town   Hong Kong   London   Madrid
Mexico City   Montreal   Munich   Paris   Singapore   Sydney   Tokyo   Toronto

Executive Editor: Leslie Berriman
Assistant Editor: Kelly Reed
Managing Editor: Deborah Cogan
Production Supervisor: Camille Herrera
Production Editor: David Novak
Copyeditor: Michelle Gossage
Compositor: Seventeenth Street Studios
Interior and Cover Designer: Seventeenth Street Studios
Cartoonist/Cover Illustrator: Kevin Opstedal
Proofreader: Laurie Dunne
Indexer: Sylvia Coates
Manufacturing Buyer: Michael Penne
Senior Marketing Manager: Neena Bali
Text and Cover Printer: Courier/Stoughton

Benjamin Cummings
is an imprint of

www.pearsonhighered.com

ISBN-10: 0-321-59592-0
ISBN-13: 978-0-321-59592-8
1 2 3 4 5 6 7 8 9 10—CRS—12 11 10 09 08

Contents

Preface

Welcome to the fascinating world of microbiology, the study of tiny, living organisms! Most students take microbiology because they have to—it's a required part of their educational curriculum. But we hope you quickly discover the amazing diversity of these microorganisms and become intent on learning all that you possibly can about them. They are important not only because they cause disease but also because they have enormously beneficial uses and effects.

If you're reading this preface, you probably have a strong desire to succeed, and you know the competition for admission into numerous educational programs is increasing. You are keenly aware that you can't just pass your classes—you need to do quality work and truly master the course content. This book offers you the opportunity to enhance your performance in this rigorous course. It is designed to help you gear up to be successful.

You may be using this book before your course officially begins. Your instructor may have assigned it to you as homework during the first week or two of class. Perhaps you are using it on your own and will come back to it periodically as you move along in your course. However you use it, the purpose of the book remains the same. The goal is to help you get a strong start in microbiology and to master the material, not just for exams, but for your future as well. *Get Ready for Microbiology* contains seven interactive chapters that engage you every step of the way. You'll read, but you'll also do activities.

The book starts with basic study skills in **Chapter 1**. As with any course, you will get out of microbiology what you put into it, and it will take significant time and effort on your part to succeed. This chapter helps you focus and manage your study time. After exploring different learning styles, you can assess which style best fits you and then discover specific study strategies that complement your preferred style. You'll assess your current habits as a student and learn specific tips and strategies to help you study better. Specific tips will help you write your notes, read textbooks, and take tests.

Chapter 2 helps you brush up on basic math skills. Microbiology is a science, and all science involves at least some math. This chapter takes you from performing basic math operations to reading and interpreting numerical information in graphs and tables—the math you'll need for a head start in microbiology.

Many of the words in your microbiology class will sound foreign to you, and well they should! Most of the terms come from Latin or Greek. A good grasp of the terminology is imperative in all aspects of learning in this class. In **Chapter 3** on terminology, we look at how the words are built and learn some simple tricks that will rapidly expand your microbiology vocabulary.

The second half of the book begins a climb up the biological hierarchy of organization that is the foundation of microbiology. In **Chapter 4**, we tackle chemistry, the first few levels of that hierarchy: atoms, molecules, and macromolecules. We discuss some tricks for gaining information from the periodic table and see how atoms join together to form molecules. If you can understand bumper cars, you can understand bonding!

In **Chapter 5**, we cover the basics of biology,

the study of life. We discuss the characteristics common to all living organisms, how living organisms are classified, and the methods scientists use to learn more about life.

Chapter 6 continues the climb up the hierarchy of organization by looking at cells, the basic unit of life. Almost everything that happens in living organisms occurs inside a cell. We first discuss the basic organization of eukaryotic cells—the cells of higher organisms. Then, we compare the features of eukaryotic cells with those of prokaryotic cells, which you will be studying in microbiology.

Finally, in **Chapter 7**, we explore the topics that you will learn in microbiology. You will learn why microbiology is so important for your career and your daily life. We also discuss strategies for learning each of the major topics in your microbiology course.

Now that you know the road map for this book, let's explore the stops you'll find along the way. Here are the special features in each chapter, designed to keep you involved and to make you a better student in microbiology:

- *Your Starting Point* pretests your grasp of the chapter content before you start. Answers are provided except in Chapter 1, where the answers are personal.

- *Quick Check* asks you to recall or apply what you just read, to keep your eyes from scanning the page while your brain is on vacation. The answer is provided on the same page.

- *Picture This* asks you to visualize scenarios and then answer questions about them to help you better understand the concepts.

- *Time to Try* is a simple experiment or quick question that gives you a chance to practice what you just learned.

- *Why Should I Care?* highlights the relevance of the material to your life and course.

- *Reality Check* asks you to apply your knowledge to the context of the real world.

- *Keys* highlight the main facts, concepts, and principles for reinforcement and easy review.

- *What Did You Learn?* end-of-chapter quizzes may include short answer, multiple-choice, or matching exercises. The answers appear at the end of the book. There is also a list of key terms from the chapter to help you start your own running vocabulary list by writing each term in a notebook or on index cards and then defining it.

Now it is time to dig in. So get comfortable and *Get Ready for Microbiology*!

Acknowledgments

Get Ready for Microbiology is adapted from *Get Ready for A&P* by Lori K. Garrett of Danville Area Community College, and from *Get Ready for Biology*, another adaptation of the original book. Lori's simple, straightforward explanations, suggestions to help students succeed, and compassion for students are evident throughout this adaptation.

Thanks go to the team that worked on *Get Ready for Microbiology*: assistant editor Kelly Reed; production supervisor Camille Herrera; David Novak; Randall Goodall and Seventeenth Street Studios; cartoonist Kevin Opstedal; and executive editor Leslie Berriman.

Finally, we owe a great debt of gratitude to all of our reviewers listed below and on the following page. They provided thoughtful and candid feedback, which helped make the book stronger and more useful to students.

REVIEWERS OF *GET READY FOR MICROBIOLOGY*

Sally McLaughlin Bauer
Hudson Valley Community College

Robert W. Bauman
Amarillo College

Judy Kaufman
Monroe Community College

Dubear Kroening
University of Wisconsin, Fox Valley

Timothy Secott
Minnesota State University, Mankato

REVIEWERS OF *GET READY FOR A&P* AND *GET READY FOR BIOLOGY*

Erin Amerman
Santa Fe Community College

Vince Austin
Bluegrass Community & Technical College

Claudie Biggers
Amarillo College

Margaret Creech
Laramie County Community College

Terry Harrison
Arapahoe Community College

Clare Hays
Metropolitan State University

Maurice Heller
El Paso Community College

Mark Hubley
Prince George's Community College

Catherine Hurlbut
Florida Community College, Jacksonville

Jody Johnson
Arapahoe Community College

Johanna Kruckeberg
Kirkwood Community College

Ken Malachowsky
Florence-Darlington Technical College

Elaine Marieb
Holyoke Community College

Judy Megaw
Indian River Community College

Claire Miller
Community College of Denver

Amy Nunnally
Front Range Community College

Wayne Seifert
Brookhaven College

Joan C. Sharp
Simon Fraser University

Alan Spindler
Brevard Community College

Dieterich Steinmetz
Portland Community College

Yong Tang
Front Range Community College

Marty Tayler
Cornell University

Deborah Temperly
Delta College

Jennifer van de Kamp
Front Range Community College

About the Authors

Lori K. Garrett is a professor in the Science Division at Danville Area Community College in Illinois. Upon her arrival at the college, she developed two human cadaver dissection courses that were the first to be offered at a two-year college in her state. She oversees the college's human cadaver program and conducts cadaver workshops for area high school anatomy students and for allied health students from schools that lack access to cadavers. She also developed a Simply Science course that prepares underprepared students for the rigors of the college's science curriculum. In this capacity, she recognized the rapidly changing nature and needs of today's students, which led to her authorship of the very popular *Get Ready* book series.

She has received numerous awards, including the National Association of Community College Trustees (ACCT) prestigious William H. Meardy Faculty Award, and an Excellence Award from the National Institute for Staff and Organizational Development (NISOD). She is a certified Leadership Development Instructor and serves as director of DACC's honors programs. She also serves on the International Honors Program Committee and the Illinois Advisory Council for Phi Theta Kappa International Honors Society. She is a frequent speaker at Human Anatomy & Physiology Society (HAPS) conferences as well as Strategies for Success workshops sponsored by Benjamin Cummings.

Judy Meier Penn is a professor of biology and microbiology and past department chair at Shoreline Community College, Shoreline, Washington. She earned a B.S. degree in biology from Concordia College (Minnesota) and an M.S. degree in botany from Louisiana State University. In her current position, she teaches microbiology, biology, anatomy, physiology, and epidemiology. She created a Biology Learning Center at the college and developed and taught special courses for English as a Second Language Students preparing for allied health careers. She authored the animations that accompany the microbiology textbooks published by Pearson Benjamin Cummings. She coauthored the textbook *Anatomy and Physiology for English Language Learners.*

Prior to her teaching career, she was a research associate in pathology and a clinical endocrinology lab manager at the University of Texas Medical Branch at Galveston. Her past research involved the ultrastructure (particularly the flagellar apparatus) and phylogeny of terrestrial algae and lichens, leprosy in armadillos, human papillomavirus infections, and parasitic infections in budgerigars. She is a member of the American Society for Microbiology (ASM) and the Northwest Biology Instructors Organization (NWBIO). When she is not teaching or writing, she enjoys traveling, gardening, reading mysteries, and spending time with her husband and cat at their island home.

Study Skills

Feeding Your Brain

When you complete this chapter, you should be able to

- Understand your preferred learning style and study strategies that emphasize it.

- Utilize skills that will help you get the most benefit from lectures, labs, and readings.

- Develop a written schedule that includes adequate study time.

- Know how to prepare well for an exam.

- Understand that you are ultimately accountable for your own success or failure.

Your Starting Point

Answer the following questions to assess your study habits.

1. How often do you read a course textbook? _____

2. Do you study hard the day before an exam but rarely between exams?

   _____

3. Where do you study?_____

4. How long should you spend studying outside of class? _____

5. Do you schedule your study time and keep the schedule? _____

6. Do you study alone or with others? _____

7. Do you mostly memorize when studying for a test? _____

8. Do you have a good support group of family and friends who encour-
   age you? _____

9. Do you quiz yourself when studying?_____

Welcome to the exciting and sometimes challenging world of micro-biology, the study of the tiniest forms of life. You will quickly discover how amazing the living world is—a curious marvel of complexity that we hope will fascinate you. Interest in your subject matter always makes it much easier to learn.

Still, no matter how exciting your explorations may be, your course may, at times, seem rigorous and demanding. You've taken a great first step by turning to this book to jump-start your studies. This book is meant to help you enter the course with a well-planned strategy for success and with confidence in your basic science knowledge. The purpose of this chapter is to help you "train your brain" to make your learning process easier and more efficient.

Why Should I Study Microbiology?

Most students take microbiology because it is required for their educational programs. Sometimes when something is required we do it only because we have to, without considering what benefits the task might hold for us. Unfortunately, some students use that approach for microbiology. Certainly it is easier to study something if you understand why it matters, and this course is no exception.

PICTURE THIS

Suppose that you are working as a nurse practitioner in a walk-in clinic. A woman arrives with a small child. She says the child has been cranky and has trouble drinking or eating because his throat hurts. She's worried and asks you what is wrong with her child. What knowledge will you need to answer the woman's question and treat her child? _____

_____

_____

Why do you need to fully understand microbiology in order to be a medical professional? _____

_____

In order to determine what is wrong with the child, the nurse practitioner will gather information about the child's health. This could include measurements such as taking the child's temperature, looking for signs of inflammation, swabbing the throat and sending that sample to the lab, maybe performing a rapid test for strep throat, and listening to the child's breathing and heartbeat with a stethoscope. She may also ask questions about the child's behavior in the past few days, how long he has been sick, whether the child has been urinating, and what the child has been eating or drinking. The nurse might want to know if the child has any siblings and whether they or other family members have been sick. The nurse would also ask whether the child has any allergies. All of these things would be important in helping the nurse practitioner determine whether the child has strep throat (a bacterial infection), a cold (a viral infection), or some other condition (perhaps even a seasonal allergy). Many of these assessments (the throat culture, inflammation, fever, allergies) reveal conditions that you will learn about in microbiology.

Now consider your own future—what are your academic and career goals? _____

_____

Why will you need to know microbiology? _____

_____

To Thine Own Self Be True: **Learning Styles**

What *is* the best way to learn microbiology? A tremendous amount of research has explored how people learn, and there are many opinions. One common approach considers which of the senses a learner relies on the most—sight, sound, or touch:

■ Visual learners learn best by *seeing.*

■ Auditory learners learn best by *hearing.*

■ Tactile (kinesthetic) learners learn best by *doing.*

TIME TO TRY

Let's uncover your learning style.

1. Look at **Table 1.1**. Read an activity in the first column; then read each of the three responses to the right of that activity.

2. Mark the response that seems most characteristic of you.

3. After doing this for each row, you are ready for your totals. Simply add all the marks in each column and write the total in the corresponding space in the bottom row.

4. Next look at your numbers. You will likely have a higher total in one column. That is your primary learning style. The second highest number is your secondary learning style.

My primary learning style is: _____

My secondary learning style is: _____

TABLE 1.1   Assessing your learning style.

Activity	Column 1	Column 2	Column 3
1. While I try to **concentrate** . . .	I grow distracted by clutter or movement, and I notice things in my visual field that other people don't.	I get distracted by sounds, and I prefer to control the amount and type of noise around me.	I become distracted by commotion, and I tend to retreat inside myself.
2. While I am **visualizing** . . .	I see vivid, detailed pictures in my thoughts.	I think in voices and sounds.	I see images in my thoughts that involve movement.
3. When I **talk to someone** . . .	I dislike listening for very long.	I enjoy listening, or I may get impatient to talk.	I gesture and use expressive movements.
4. When I **contact people** . . .	I prefer face-to-face meetings.	I prefer speaking by telephone for intense conversations.	I prefer to interact while walking or participating in some activity.

▶

TABLE 1.1 Assessing your learning style, continued.

Activity	Column 1	Column 2	Column 3
5. When I **see an acquaintance** . . .	I tend to forget names but usually remember faces, and I can usually remember where we met.	I tend to remember people's names and can usually remember what we discussed.	I tend to remember what we did together and may almost "feel" our time together.
6. When I am **relaxing** . . .	I prefer to watch TV, see a play, or go to a movie.	I prefer to listen to the radio, play music, read, or talk with a friend.	I prefer to play sports, make crafts, or build something with my hands.
7. While I am **reading** . . .	I like descriptive scenes and may pause to imagine the action.	I enjoy the dialogue most and can "hear" the characters talking.	I prefer action stories, but I rarely read for pleasure.
8. When I am **spelling** . . .	I try to see the word in my mind or imagine what it would look like on paper.	I sound out the word, sometimes aloud, and tend to recall rules about letter order.	I get a feel for the word by writing it out or pretending to type it.
9. When I **do something new** . . .	I seek out demonstrations, pictures, or diagrams.	I like verbal and written instructions and talking it over with someone else.	I prefer to jump right in to try it, and I will keep trying and try different ways.

TABLE 1.1 Assessing your learning style, continued.

Activity	Column 1	Column 2	Column 3
10. When I **assemble something** . . .	I look at the picture first and then, maybe, read the directions.	I like to read the directions, or I talk aloud as I work.	I usually ignore the directions and figure it out as I go along.
11. When I am **interpreting someone's mood** . . .	I mostly look at his or her facial expressions.	I listen to the tone of the voice.	I watch body language.
12. When I **teach others how to do something** . . .	I prefer to show them how to do it.	I prefer to tell them or write out how to do it.	I demonstrate how it is done and ask them to try.
TOTAL:	**Visual:** _____	**Auditory:** _____	**Tactile/Kinesthetic:** _____

(Source: Courtesy of Marcia L. Conner, www.agelesslearner.com)

Now that you know your primary and secondary learning styles, you can design your study approach accordingly, emphasizing activities that use your preferred senses. Look closely at your scores, though. If two scores are rather close, you already use two learning styles well and will benefit from using both of them when studying. If your high score is much higher than your other scores, you have a strong preference and should particularly emphasize that style. Most people use a combination of learning styles.

In addition, information coming in through different senses reaches different parts of your brain. The more of your brain that is engaged in the learning process, the more effective your learning will be, so try strategies for all three styles and merely emphasize your preferred style over the others. You'll know which strategies work best for you. We'll consider some strategies that you might try for each style; these ideas are summarized for you in **Table 1.2**.

TABLE 1.2   The three learning styles and helpful techniques to use in your studies.

 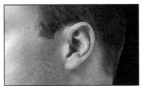 

	Visual	Auditory	Tactile
Techniques to use	❏ Sit close to the instructor. ❏ Take detailed notes. ❏ Draw pictures. ❏ Make flowcharts and concept maps. ❏ Use flash cards. ❏ Focus on the figures, tables, and their captions. ❏ Try coloring books and picture atlases. ❏ Use visualization.	❏ Listen carefully to your instructor's voice. ❏ Read the textbook and your notes out loud. ❏ Record lectures and listen to them later. ❏ Listen during class instead of writing notes. ❏ Work in a study group. ❏ Discuss the material with others.	❏ Highlight important information while reading. ❏ Write your own notes in class and while reading the textbook. ❏ Transfer your notes to another tablet or type into your computer. ❏ Doodle and draw as you read. ❏ Build models of biological structures, using clay or other material. ❏ Create and conduct your own experiments. ❏ Hold your book while reading. ❏ Walk or stand while reading. ❏ Use microbiology coloring books. ❏ Make and use flash cards.

VISUAL LEARNERS

If you are a **visual learner**, you rely heavily on visual cues. You notice your instructor's mannerisms, expressions, gestures, and body language. Seeing these cues is especially helpful, so sit at the front of the classroom, close to the instructor. You tend to think in pictures and learn well from visual aids such as diagrams, illustrations, tables, videos, and handouts. Table 1.2 has some strategies for you.

AUDITORY LEARNERS

If you are an **auditory learner**, you learn well from traditional lectures and discussion. You listen carefully to your instructor's vocal pitch, tone, speed, and mannerisms. Material that you struggle with while reading becomes clearer when you hear it. See Table 1.2 for some specific strategies.

TACTILE LEARNERS

If you are a **tactile learner**, you learn best by actively participating and doing hands-on activities. You may become bored easily in class from sitting still too long and start fidgeting or doodling. You need to do something physical while studying and learning. Table 1.2 has some strategies for you.

WHY SHOULD I CARE?

Understanding your own learning style allows you to develop more effective and efficient study techniques that take advantage of your sensory preferences. By emphasizing your preferred learning style, the material will be easier to learn and will stay with you longer.

 **QUICK CHECK**

Homemade flash cards would be most beneficial to which two learning styles? _____ and _____.
How could they be used to benefit a learner of the third style?

_____

Answer: They would benefit visual and tactile learners. Reading them out loud would benefit auditory learners.

Putting Your Best Foot Forward: **Getting Ready**

Many students mistakenly wait until the first lecture to start thinking about class. The key to starting your semester well is to be organized and ready when you enter the classroom. This requires you to plan ahead, but the time invested will save you tremendous time while the semester is underway.

PUTTING IT IN WRITING

As the semester begins—if not before—you should get organized, and that begins with making a commitment to yourself. Too often we begin a project without setting goals in advance. If you set a goal, you enter with a purpose and a direction. If you do not set a goal, it's too easy to just go along and see where you end. Take time to think about your goals for the semester. They should be both specific and attainable. Be realistic. For example, it may not be realistic to set the goal of always having the highest score in class, but a goal of getting an A in the class might be attainable. Once you've decided on your goals, write them down to give them more importance. Once you've written them, be firmly committed to them. To reinforce these goals, write them on an index card and place it in a prominent location in your study area so you'll see them every day.

TIME TO TRY

Set three main goals for yourself in this class, and write them below. Explain why achieving each goal is important to you.

*Goal 1:*_____

*It is important to me because:*_____

_____

*Goal 2:*_____

*It is important to me because:*_____

_____

*Goal 3:*_____

*It is important to me because:*_____

_____

PULLING IT ALL TOGETHER

Have you ever seen a fellow student show up for a test with no pen or pencil? Don't let that happen to you. The more organized you are, the more efficient you will be, so let's organize what you will need for class. Categorize the items by what you take to class every day, what remains at home in your study spot, and optional items that are nice, but nonessential, additions. Use the checklist provided for you in **Table 1.3**.

You need a pocket-sized day planner that has plenty of room for writing and that you can keep with you at all times. Or you may opt for a personal organizer portfolio or an electronic organizer. Select one you like because you'll use it every day. In it, write all important dates you already know—when classes begin, holidays, last day to withdraw from a class, final exam schedule. Enter all class times, your work schedule, and any other known time commitments. Try to keep your day planner current so you always know how your time is being spent and can plan ahead.

If it's not part of your day planner, you need a separate to-do list. You will write all assignments and due dates on this list. You want one single to-do list for all of your classes as well as nonschool activities because they must all be done from the same pool of time. Writing them down allows you to view the entire list and review the deadlines for each item so you can easily prioritize, doing the assignments in the order in which they are due.

Maintain a record of all grades you receive (**Figure 1.1**). For each graded item, list what it is, when you turned it in, when you got it back, how many points you received, how many points were possible, and any additional notes. Once you know how your grade will be determined for the course, you can use this to keep track as you go along. It also provides a backup in case there is any confusion later about your grade or your work.

Check with your instructor to see if you should bring your textbook to class. Typically, you may not need it in lecture, but you may in lab or vice versa. If your course uses a lab manual, always take it to lab.

Set up your home study space like a home office. Be sure to have all the essential office supplies on hand—plenty of writing utensils, paper, a stapler, and so on. A critical part of the home study area is the master calendar. There are large desktop versions and wall charts, for example.

TABLE 1.3 Organizer's checklist.	
Item	✎✗
To take to class each day:	
Book bag/backpack/rolling carrier	
Textbooks	
Lab manuals	
Pocket-sized day planner	
To-do list	
Separate notebooks for each course	
Copy of class schedule with buildings and room numbers	
Several blue or black ink pens	
Several #2 pencils	
Small pencil sharpener	
2–3 colored highlighter pens	
Small stapler	
Grade record sheet for each course	
Calculator	
Computer	
At home:	
Master calendar	
Separate file or folder for each course	
Loose notebook paper	
Index cards for making flash cards	
Computer paper (if I have a computer)	
More writing utensils (pens and pencils)	
Stapler	
Calculator	
Scissors	
Paper clips	
Computer	
Optional:	
Personal organizer	
Microbiology coloring book	
Colored markers/pencils	
Small recorder to record lectures/readings	
Extra batteries	
Biological dictionary	

Graded item	Date turned in	Date returned	My score	Possible points	Notes
Lab 1	9/6	9/13	10	10	Worked with Emily, Mike, Tom
Quiz 1	9/7	9/9	18	20	Study terms again
Lab 2	9/13	9/16	6	10	Messed up the math!
Pop Quiz	9/14	9/16	5	5	From yesterday's lecture.
					I was ready!
Quiz 2	9/21	9/25	19	20	Forgot to answer one question!

FIGURE 1.1 A sample grade record for keeping track of your progress.

You could use a calendar feature on your computer, but the more visible the calendar is, the more often you will look at it. This calendar should be large enough to accommodate plenty of writing, so think BIG! Each day, you should add anything that you put in your day planner or on your to-do list to this master calendar. All time commitments should be entered, so also add all personal appointments and vacations. This is how you will schedule your life while in school, and the practice will likely stick with you far beyond that.

✔ **QUICK CHECK**

To be successful in class, your effort should start before class begins. What are some tasks you should do before the first day of class?

_____

_____

_____

Answer: Set and write down your goals, organize the items you will need for class and at home, pack your carrier, start your day planner and master calendar, and organize your study space.

I Hate to **Lecture** on This, but Can You Hear Me Now?

Welcome to class! Imagine that it is the first day. You walk into class.

Where do you sit? _____

Why do you sit there? _____

The best seat in the house is front and center. Obviously not everyone can sit there, but you should arrive early enough to sit within the first few rows and as near to the middle as possible. You want an unobstructed view of the instructor and anything he or she might show because microbiology is often a visual course. People sitting on the sides or in the back often do not want to be called on, or they want to be in their own space. They are not very engaged in the class. Don't let that be you. To succeed, you need to focus all of your attention on your instructor, minimize distractions, and actively participate. Instructors tend to teach to the middle of the room. In fact, if your instructor is right-handed and uses equipment, such as an overhead projector, that is positioned on the right, the instructor's focus shifts to his or her right. You want to see your instructor, and you want your instructor to see that you are present, actively listening, and engaged.

Some instructors provide lecture notes so you can sit back and really think about what is being said. Notes or not, you need to get all the information you can from each lecture. Remember your learning style and use techniques that enhance it. We will discuss note taking momentarily, but consider recording the lectures (ask permission first). That way you miss nothing, and you can listen to the lecture repeatedly, rewinding as needed. Another good technique is to write out your own notes while listening to the recording and then listen again while reading your notes and making necessary corrections. This combination strongly reinforces the material.

Always try to preview the material that will be covered before going to class. This is as simple as lightly reading the corresponding sections in the textbook. You may not understand all that you read, but it will sound familiar and be easier to comprehend as your instructor covers it in class. This preview also helps you identify new vocabulary words.

While your instructor is lecturing, don't hesitate to raise your hand to ask a question or get clarification. Many students are shy and reluctant to speak in class—you may be doing them a favor! Avoid discussing personal issues in front of the whole class—that is better done alone with the instructor, outside the classroom.

Note your instructor's gestures, facial expressions, and voice tone for clues about what your instructor finds most important. That material is likely to show up on a quiz or test. Write down any material that is particularly emphasized, or mark it in your notes. Listen carefully for assignments and write them down immediately on your to-do list. If you are not clear about the expectations of the assignment or when it is due, seek immediate clarification.

✔ **QUICK CHECK**

Why is it best to sit front and center in class?

_____

_____

Answer: You will be more engaged in the class, have the best view and fewer distractions, and be within your instructor's focal area.

Taking Notes: It's Not Brain Surgery

Anybody can take notes in class, but will the notes be good enough to help a student succeed? There are many strategies and models for how to take notes, and none of them is the best. Find what works for you and then use it consistently. Let's review one easy-to-use system (**Figure 1.2**).

Start with a full-sized (8.5" × 11") notebook that you will use just for this class. Take your notes on only the front side of the paper and leave about a 2" margin on the left. The margin will be used for marking key words and concepts later. At the beginning of class, date the top of the page so you know when the material was covered. During lecture, use an outline format to get as much information down as you can. Use the main concept as a major heading; then indent the information discussed on that topic. When that section ends, either draw a horizontal line to mark its end or leave a couple of blank lines. Don't try to write every word—just the main ideas—and put them in your own words. Instead of writing out every example, give a brief summary or a one- or two-word reminder. Use abbreviations when possible, and develop your own shorthand. You can often drop most of the vowels in a word and

03/14/06

RECALL:

Note
taking

Reviewing

I.   Note-taking tips

    A. Use outline format

    B. Be concise

    C. Get main ideas

II.   Reviewing notes

    A. Review after class

        1. Fill in gaps

        2. Clean up

        3. Replay lecture in my mind

        4. Review within 24 hours—fresh in mind.

3 learning
styles

III.   Learning styles

    A. Visual—reread and add drawings

    B. Tactile—rewrite or type

    C. Auditory—read out loud or tape-record

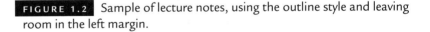

FIGURE 1.2   Sample of lecture notes, using the outline style and leaving room in the left margin.

still be able to sound it out later when reading it. Write legibly or your efforts will be useless later. Underline new or stressed terms and place a star or an arrow by anything that is emphasized. Be as thorough as you can, but you will need to write quickly. The instructor will not wait for you to catch up, so speed is essential. If your instructor makes Power-Point lecture notes available, print them in a three-slides-per-page format and use these as your note taking templates. You can also bring them in your laptop and take notes right on the computer.

As soon as possible after class, read your notes and improve them as necessary. Add anything that's missing. Make them clearer and cleaner. Put the concepts in your own words. Next, use the left margin to summarize each section—the main concept, subtopics, and key terms. The latter column will be your "Recall" column. Once you are sure all of the key ideas are in the left column, you can cover the right side of the page—the meat of your notes—and quiz yourself on the main points listed in the left column. It makes an easy way to review.

But you are not finished—if you are a tactile learner, rewrite your notes in another notebook or type them into your computer. Visual learners might type and reorganize the notes. Auditory learners can read the notes out loud or record them. You can make flash cards from the key points and terms by writing the term on one side of an index card and its definition or use on the other side. You can add drawings. Review your notes as much as you can during the next 24 hours, while the lecture is still fresh in your mind.

TIME TO TRY

Look at the sample notes in Figure 1.2. Now practice: Take notes using this style while listening to a 1-hour TV show. Capture the conversations and action in words. You can't get every word down, so paraphrase—put it in your own words so the meaning still comes across. When you're finished, assess how you did.

Can you tell who was talking? _____

Do your notes make sense? _____

Did you capture the main ideas? _____

Did you keep up or fall behind? _____

Do you have breaks in your notes to separate the main conversations and action? _____

What can you do better when taking notes in class?

_____

✔ **QUICK CHECK**

What should you do with your notes after class? _____

_____

_____

Answer: Review them within 24 hours, fill in anything missing, clean them up, put them into your own words, add key concepts and terms to the Recall column, add drawings, make flash cards, record them, and rewrite or type them.

SQRHuh? **How to Read a Textbook**

The name may sound odd, but **SQR3** is an effective method for studying your textbook. Science textbooks do not read like novels, so you need to approach them differently. This method also works well for reviewing your notes. It stands for

- Survey

- Question

- Read

- Recite

- Review

During the *survey phase,* read the chapter title, the chapter introduction, any other items at the beginning of the chapter, and all of the headings. This gives you the road map of where you will be going in the chapter. As you skim the chapter, also read all items in bold or italic. Next, read the chapter summary at the end of the chapter.

During the *question phase,* look at the heading of each section and form as many questions as you can that you think may be covered in that section. Write them down. Try to be comprehensive in this step. By having these questions in mind, you will automatically search for answers as you read.

Now *read* the chapter for details. Take your time. Adjust your reading speed with the difficulty of the material. Also, keep in mind the questions you developed and try to answer them.

The next phase is to *recite*. You are working on your ability to recall information. After reading each section, think about your questions and try to answer them from recall. If you cannot, reread the section and try again. Continue this cycle until you can recite the answers, out loud if necessary.

Finally, you want to *review*. This helps reinforce your memory. After you complete the previous steps for the sections you're studying, go back to each heading and see if you can still answer all of your questions. Repeat the recite phase until you can. When you are done, be sure you can also answer the questions at the end of the chapter and on the Web site.

✔ **QUICK CHECK**

What is SQR3 and what do the letters mean?

_____

_____

Answer: It is a system for reading a textbook, and the letters mean Survey, Question, Read, Recite, and Review.

Cooking Up Some Fun **in the Lab**

Your microbiology class comes in two parts: lecture and lab. Many students put most of their efforts into the lecture material and disregard the lab component. Avoid this. Lab is the hands-on part of the course, and most people learn better by seeing demonstrations and actually doing the work themselves. Also, part of your grade comes from your performance in lab. Always go to lab prepared to take notes, equipped with your lab manual, if required, and your textbook, if it will be needed.

When in lab, you may work with a lab partner or group. You will be expected to contribute equally to the team effort, so it is important that you arrive prepared for lab. If you know in advance what the lab will be, read through it and think about what you will be doing. Pay attention to the instructions and especially note any safety precautions. Many microbiology lab courses contain an "unknown identification" activity in which you will be given a bacterial species unknown to you and be required to perform a series of tests to identify it. You will be the detective, searching methodically for the identity of this unknown and gradually eliminating "suspects" as you perform each test. Most students enjoy this activity because it's an experience much like that of a clinical microbiologist, searching for the agent that's causing a disease.

Your instructor may require that you keep a lab notebook that documents this process. Be sure you understand how your instructor wants this notebook formatted and how much information you are to include. You should always write in pen, cross out errors with a single line (never use correction fluid or erasers), and include as much detail as possible about your procedures and observations. A lab notebook is a working document that is recorded as you work, never after the fact. So, don't worry if you make a few mistakes. The key to doing well in this project is to understand and follow the directions, plus be able to interpret results correctly.

Some students try to take shortcuts in lab so they can leave a bit early. You should value lab as a time to further explore the material covered in lectures. It is a unique aspect of your education that reinforces everything else that you are learning. The more time you spend in lab, the more you learn. Remember that lecture and lab are both part of the same class, and try to see how they fit together. Never leave lab early—there's always more to learn.

PICTURE THIS

You are working in a group of four in the lab. One of your group members is not participating but just watching. What could be some possible reasons why the person is watching and not helping?

_____

Perhaps this person hasn't prepared. Perhaps this student lacks self-confidence and is afraid he or she will mess up the experiment for the rest of the group. In the latter case, how could you help this student gain confidence? _____

_____

You could suggest that he or she do a particular part of the experiment that is easiest. Or, you could ask the student to help you to do a particular portion of the work. Sometimes people feel more comfortable if they are helping someone else, rather than working independently. How would you respond to a team member who insists on doing everything himself or herself and won't let you participate?

_____

Try to imagine why the student is acting that way. Could it be because he or she is concerned about the grade and wants no one else to influence it? Could it be because he or she is not sure about the abilities of others in the group and whether other students can be trusted to do it well? Your goal is to reassure that person that you are capable. You can do this best by being prepared for lab and saying something like "I'd really love to try that. Could you watch me to see if I'm doing it right?" Why do you think it's important for your future career to be able to work well with all sorts of people?

_____

 The more time you spend in lab, the better you will learn. ▪

Your Secret Life **Outside of Class**

REALITY CHECK

Answer True or False to each of the following statements:

1. I study the day before a test but rarely study on a daily basis.

   T   F

2. I mostly review my notes and don't read the textbook.

   T   F

3. I am too busy to study each day.

   T   F

4. When I finally get around to it, I study pretty hard for a long time.

   T   F

5. I get by fine with cramming and often do an "all-nighter."

   T   F

JUST FOR FUN

Let's see how well you REALLY study! Take a few moments to study and learn these terms. We will come back to this exercise a bit later.

1. **Frizzled greep.** This is a member of the *Teroplicanis domesticus* family with girdish jugwumps and white frizzles.

2. **Gleendoggled frinlap.** This is a relatively large fernmeiker blib found only in sproingy sugnipers.

3. **Borky-globed dungwinger.** This groobler has gallerific phroonts and is the size of a pygmy wernocked frit.

Stay tuned!

You made it through lecture or lab and are ready to head home. Finally! School is done for the day, right? Not if you plan to be successful! The real work begins after class because most of your learning occurs outside the classroom on your own. This is often the hardest part, for many reasons. We schedule many activities and set aside time for them, but studying tends to get crammed into the cracks. Too often, studying becomes what you do when you "get around to it." It is an obligation that often gets crowded out by other daily activities and the first item dropped from the to-do list.

Too many students only study when they have to—before a quiz or exam. A successful student studies every day. The goal is to learn the material as you go rather than frantically try to memorize a large amount at the last minute. Here is something you need to know and really take to heart.

You should study for at least 2 to 3 hours for every hour spent in class. ▪

Simple math shows you that if you have three lectures on Monday, for example, you should plan to spend from 6 to 9 hours studying that same day. YIKES!

SCHEDULE YOUR STUDY TIME

Writing assignments on your to-do list makes them seem more urgent, but that does not cover the daily work that must be done. You must take charge of your time and studying. In addition to specific assignments, each day you should

1. Go over that day's notes.

2. Read the corresponding sections in the textbook.

3. Quiz yourself.

4. Review your notes again.

5. Preview the next day's material.

All of this takes time. You must build study time into your schedule or you either will not get around to it or you will put it off until you are too tired to study effectively. The first thing to do is to write your study time into your day planner and master calendar and regard that time as sacred—do not borrow from it to do something else. Be sure to allow break time during study sessions as well—if you study for too long, your brain gets weary and starts to wander, and it takes much longer to do even simple tasks. Plan a 10- to 15-minute break for every hour of studying.

CHUNK IT

If a job seems too large, we put it off, but if we have many small tasks, each alone seems manageable. Break your workload into small chunks. Write them down, partly so you do not forget any, but especially because you will get a great feeling of accomplishment when you complete a task and cross it off your to-do list! Completing a task is also a convenient time to take a mini-break to keep your mind fresh. Many students try to read a whole chapter or cover a few weeks of notes in one sitting. The brain really dislikes that. When studying a large amount of material, divide it into subcategories and then study one until you really understand it before moving to the next.

STUDY ACTIVELY

Merely reading your notes or the book is not learning. You must think about the material and become an **active learner**. Constantly ask yourself, "What is most important in this section?" While reading, take notes or underline key terms and major concepts. Make flash cards. After reading a section, close the book and ask yourself, "What did I just read?" If you have no idea, you have not learned anything and should try again. Consider how what you are studying relates to something with which you are already familiar. If you can put the information in a familiar context, you will retain it better.

The best preparation for quizzes and tests is practice. Develop and answer questions as you read. Try to anticipate all the ways your instructor might quiz you about that material. Recall which specific items your instructor stressed. Outline the material in each section and be sure to understand how the different concepts are related. Check yourself on the meanings of the key terms. Say the key words out loud and look carefully at them. Do they remind you of anything? Have you heard them before?

MOVE PAST MEMORIZING

This is one of the hardest study traps to avoid. In microbiology, it may at times seem like there is so much to learn and so little time. Most students at first attempt to just memorize. If you only read your notes and the book, you are using this approach without realizing it.

At the beginning of this section, we gave you three items to learn. Without turning back, write down the three names you were asked to learn a few pages ago:

1. _____

2. _____

3. _____

Did you remember them? Now, also without looking back, can you explain each of them? _____

(Probably not.)

These three "things" are fictitious, but the point is that you may indeed have memorized the names—it doesn't take much to memorize—but it takes a lot more to understand, especially if the words are unfamiliar, as they often are in microbiology. If you find that you study hard but the wording of the quiz or test confuses you, you are probably memorizing. The question is worded a bit differently than what you memorized so you don't realize that you know the answer. You must get past memorizing by looking for relationships between the concepts and terms and really strive for full understanding. Reading often produces memorization. Active studying produces understanding.

THE CONCEPT MAP

A useful technique for learning relationships is drawing a **concept map**. This is somewhat like brainstorming. Here is the general process:

1. Start with a blank piece of (preferably) unlined paper.

2. Near the center, draw a circle and, inside it, list the main concept you will explore.

3. Around that circle, and allowing some space, draw more circles and list in each anything that pops into your mind related to your main concept. Do this quickly and don't think about the relationships yet. Just get your ideas down.

4. Once you've added all your secondary concepts, look at them and think about how they are related, not just to the main concept but to each other as well.

5. As relationships occur to you, draw arrows connecting related concepts and add a brief description of the relationship between each of the concepts.

6. Examine the relationships and you will start to understand how these concepts fit together.

Another way to make a concept map is to write key terms on index cards and arrange them to form a concept map.

TIME TO TRY

Construct a concept map around the main concept of *energy* by adding arrows to show relationships between the following concepts:

- Cell activity

- Food

- Plants

- Sun

- Work

When you are finished, look at **Figure 1.3** on page 27.

CRITICAL THINKING

You've learned that memorizing is only a part of learning. You also need to make connections between various topics in your course. You must use that information to solve problems. This is all part of a process called **critical thinking**.

Recall, from the beginning of this chapter, the story about the nurse practitioner in the walk-in clinic. She gathered information about a sick child who had a sore throat. Then she sorted through it to determine what was relevant. The fact that the child ate cornflakes for breakfast may not be relevant. However, the fact that he is not able to swallow might be important.

How does she decide what is relevant? The nurse has learned the signs and symptoms of diseases such as colds, strep throat, and other respiratory infections. She knows how they differ (not all exhibit fever or inability to swallow) and what the various test results mean. She's sorted through the information and has a short list of important features that lead her to a diagnosis of strep throat.

Then she must decide whether to treat the disease with an antibiotic or tell the mother to just let the child rest. The answer will be deter-

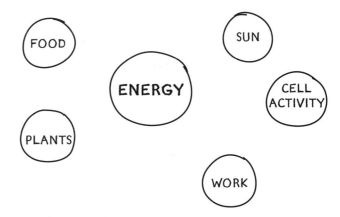

First draw a circle or "node" for each concept, keeping the main concept, if there is one, near the middle.

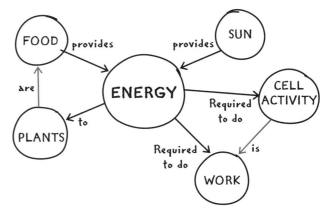

Next, add arrows linking the different concepts to each other and then add brief descriptions of how they are linked. In this example, the arrows show that the sun provides energy to the plants, and plants are food that provides energy that is required to do cell activity, which is a type of work.

FIGURE 1.3   Drawing a concept map.

mined by additional information: She knows the infection is bacterial (and antibiotics work on bacterial infections), she knows the child has no known medication allergies (so she can prescribe the antibiotic that's been most effective), and she knows the child's age and general health (so she knows what antibiotics are not appropriate for that age). Now she can decide what to prescribe.

It is not enough to just memorize information. The process of critical thinking involves determining what information is relevant, knowing the connections between certain concepts, and making choices based on that information. ■

REVIEW

OK, you've been at this study thing for awhile and you think you're starting to get the material. You did all of the prescribed steps; can you quit now? Almost, but once you think you have the material under control, review it one more time. Repetition is the key to your long-term memory—the more you go over the material, the longer it will stay with you. We always recommend a minimum of three passes, even for easy stuff—read your notes, read the text, reread your notes—and that assumes you are understanding the material. Or, do this twice on your own and then a third time with your study group. Always slow down and go through it more if you are struggling with a certain section. Use **active learning** with each pass and then finish with one more review. If you are alert enough to review right before bed, once you are asleep your brain often continues going over the material without bothering you too much.

The Web site for your textbook also provides a good way to review. The address is in your book. Most book Web sites offer a wide variety of activity options, perhaps including online animations, flash cards, puzzles, objectives, vocabulary lists, and quizzes. When taking online quizzes, be sure to do so without looking in your book or your notes. Not using them provides a better simulation of the classroom experience. Also, if you do well on the online quizzes without using your notes, you will have confidence in the classroom.

NO CRAMMING ALLOWED!

You probably have a busy life so your home occasionally gets a bit cluttered. If company drops by and it is a bad day, you might quickly grab some of the clutter and cram it into a spare closet. After the guests leave, you open the closet door to pull out a quilt. What happens?

_____

_____

Now imagine what you do to your brain when you cram for an exam. You are essentially opening the closet door, cramming stuff in, and slamming the door. When you are taking the test, you open the door to pull out the answer you need, but anything might tumble onto your paper. Cramming at best allows partial memorization. At worst, it causes the information to get mixed up and you fail. It is a desperate act of superficial studying guaranteed to NOT get you through microbiology. If you study on a daily basis instead of doing a panicky cram session before a test, you will be calmly reviewing what you already learned well and smiling at the crammers in class.

NO VAMPIRES ALLOWED!

Do you think you can pull an all-nighter and really do well?_____

What do you think are some of the reasons this will not work?_____

_____

If you normally live your life by day, you cannot suddenly override your natural biological clock and expect your brain to stay alert and focused when it knows it is supposed to be asleep. Caffeine may help keep your eyes open, but you'll only be a tad more alert, as well as jittery while still yawning and mentally drifting away from the task at hand. The only reason you would need to stay up all night is if you haven't been studying all along and this is your last option. It is not effective. You will not be mentally alert. You will not be able to focus or think through the material. Your eyes may skim the pages, but you'll struggle to comprehend the words, and you'll retain only a tiny amount of the very little that you absorb.

An all-nighter is basically a marathon cram session held at the worst possible time. It simultaneously robs your brain and body of what they need—restoration before the next day. You may be able to stay awake all night, but if you doze off you may oversleep and miss your exam. Or, if you do arrive, you may get partway into the test only to have your brain bail on you. If you are prone to "test anxiety," your defenses will be down, and you will quite likely freeze and fail. Ah, if only you had been studying all along . . .

For your brain to be kind to you, you have to be kind to its home. You must take care of yourself physically—eat, sleep, exercise, and RELAX. ■

✔ **QUICK CHECK**

Why should you study every day if the test is not for 2 weeks?

_____

_____

Answer: Studying on a regular basis breaks the material into smaller, more manageable pieces that you can master; the material is fresh in your mind, and you will only need to review it before the test.

Strength in Numbers: **The Study Group**

One of the best ways to learn anything is to teach it to someone else, so form a study group or discuss the material with others around you (**Figure 1.4**). Although this may not be the best option for everyone, it is highly effective for many students. As soon as possible, start asking your classmates who wants to be in a study group—you *will* get people to join. You can quiz each other, discuss the material, help each other, and quite importantly, support each other. If you study solo, you may not be aware of your weaknesses. Your study partners can help identify them and help you overcome them. A good way to work in a study group is to split up the material and assign different sections to different members, who then master the material and teach it to the group. Each member should also be studying it all on his or her own—that ensures better effort from everyone and allows other members to correct any errors in a presentation.

Scheduling joint study sessions can be challenging. Many students find that scheduling group sessions before or after class works best. You may want to establish some ground rules, including agreeing to use the time for studying and not for gossiping or just socializing. And although it may be tempting to meet over a pizza, you do want a quiet location

Study groups can be helpful for staying motivated and focused on course material.

where you can freely discuss the material with few distractions. Check with your instructor to see if there are any open lab times that might work for this, or see if your college library has group study rooms you can reserve.

A Place to Call My Own: **The Study Environment**

Briefly describe the location where you plan to do most of your studying.

_____

_____

_____

Now you know how to study effectively, but we often overlook WHERE to study. Your options may be limited, so you need to make the best of what you have. Ideally your study spot is somewhat isolated and free of distractions such as TV, music, and people. At the least, you should minimize distractions.

Do you study in front of a TV that is on? Even if you try hard to ignore it, you will be drawn to it, especially if the material you are studying is tough. Music can be tricky—choose music that keeps you calm and more focused.

Thinking about the study site you just identified, what distractions might you face? _____

_____

How can you minimize them? _____

_____

Your study spot should be quiet, and it should have good lighting to avoid eyestrain, a comfortable chair, good ventilation and temperature, and a work surface on which you can spread out your materials. Ideally it will be a place where you do nothing but study, so that when you are seated there you know exactly what your purpose is. If you are having trouble staying on task in your work spot, get up and walk away briefly. The mental and physical break may help you "come back" to work, and you won't begin associating the spot with struggling.

You will spend a lot of time studying here, so take time initially to set up your study space. The area should be uncluttered and well organized. It should also be inspiring and motivational. Perhaps you could frame a list of the goals you set at the beginning of this chapter, and display them boldly and prominently. Consider displaying a photo of your hero or an inspirational quote. With these treasures surrounding you, you're a mere glance away from being reinvigorated if a study session starts to fizzle.

WHY SHOULD I CARE?

Most of your learning is done outside of class. The more efficient your studying, the better you will learn. Your study spot affects your attitude and concentration. The more seriously you take your study location, the more seriously you will study there.

If you live with family or roommates, you absolutely must stress to them the importance of respecting your study time and study space. Be sure they know your academic goals and why they are important to you, and ask them to help by giving you the time and space you need to succeed. Ask them not to disturb you when you are in your space. If you have too many distractions at home, the solution is to study elsewhere. Whether on campus, in the local library, or at a friend's house, you need a distraction-free setting, and if you can't get it at home, remove yourself instead of trying to cope with a poor study space.

✔ **QUICK CHECK**

What are some of the main considerations in selecting your study spot? _____

_____

Answer: Few distractions, own space just for studying, comfortable, good lighting/ventilation, sufficient work space, and welcoming environment.

My, How **Time** Flies!

You know you need to study and that it takes a lot of time, but how will you fit it in? Let's discuss a few ways to budget time for studying. First, be consistent. Consider your schedule to see if you can study at the same time each day. Studying will become a habit more easily if you always do it at the same time. Some students adhere to one schedule on weekdays and a different one on weekends. When scheduling study time, consider your other obligations and how distracted you might be by other people's activities at those times. Don't overlook free hours you might have while on campus. Head to the library, study room, or a quiet corner. This is the ideal time to preview for the next class or to review what was just covered.

TIME TO TRY

This is a two-part exercise designed to help you find your study time.

Part A: Each week has a total of 168 hours. How do you spend *yours?* **Table 1.4** on page 35 allows you to quickly approximate how you spend your time each week.

1. Complete the assessment in Table 1.4 to see how many hours are left each week for you to study.

2. Enter that number here. _____ hours

Part B: Next, turn your attention to **Table 1.5** on page 36.

1. Enter your class schedule, work schedule, and any other activities in which you regularly participate.

2. Now look for times when you can schedule study sessions and write them in.

3. Are you able to schedule 2 to 3 hours of study time per hour of class time? _____

It can be difficult, but it is essential to make the time. Writing it into your schedule makes it more likely to happen.

Don't overbook! Be sure to build in break time during and between your study sessions, especially the longer ones. Allow for flexibility—realize that unexpected events occur, so be sure you have some extra time available. Also, be sure you plan for and schedule recreation, too. You cannot and should not study all the time, but these other activities do take time and need to be in your schedule as well so that you do not double-book yourself.

Putting It to **the Test**

Some people suffer from true test anxiety, but the majority of students who claim to have this condition believe it to be true not because of an actual diagnosis, but rather because they get very nervous and may go blank during tests. If a class is asked who amongst them suffers from test anxiety, most hands go up. By the end of the semester, with some coaching, the number is far less. Why? They have learned how to take tests and how to stay calm. If you do suffer from true test anxiety, consult with your instructor right away so he or she can put you in contact with the support services you need to understand your condition and to learn how to conquer it.

TABLE 1.4   Assessing how your time is spent. For each item in this inventory, think before answering and be as honest as possible. Items that are done each day must be multiplied by 7 to get your weekly total. One item may be done any number of times a week, so you'll need to multiply that item by the number of times each week you do it. After you have responded to all the questions, you'll have an opportunity to see how many hours remain during the week for studying.

Where does your time go? Record the number of hours you spend:	How many hours per day?	How many days per week?	Total hours per week: (hours × days)
1. **Grooming,** including showering, shaving, dressing, putting on makeup, and so on.			
2. **Dining,** including preparing food, eating, and cleaning up.			
3. **Commuting** to and from class and work, from door to door.			
4. **Working** at your place of employment.			
5. **Attending class.**			
6. **Doing chores** at home, including housework, mowing, laundry, and so on.			
7. **Caring** for family, a loved one, or a pet.			
8. **Doing extracurricular activities** such as clubs, church, volunteer work.			
9. **Doing errands.**			
10. **Participating in solo recreation,** including watching TV, surfing the Internet, reading, playing games, working out, and so on.			
11. **Socializing,** including going to parties, talking on the phone, hanging out, dating, and so on.			
12. **Sleeping** (don't forget those naps!).			
Now add all numbers in the far column to get the total time you spend on all these activities.			
		Hours/week	168
		Total hours spent on other activities −	
		Left for studying =	

TABLE 1.5 My study schedule.

Time	Sun.	Mon.	Tues.	Wed.	Thurs.	Fri.	Sat.
6:00 AM							
7:00 AM							
8:00 AM							
9:00 AM							
10:00 AM							
11:00 AM							
Noon							
1:00 PM							
2:00 PM							
3:00 PM							
4:00 PM							
5:00 PM							
6:00 PM							
7:00 PM							
8:00 PM							
9:00 PM							
10:00 PM							
11:00 PM							
Midnight							

Most people dread taking tests and experience some anxiety when taking them. Not surprisingly, the better prepared you are for an exam, the less worried you will be. The best remedy for the stress you associate with taking tests is to be well prepared. If you know you understand the material, what is left to worry about?

Some people get anxious before tests because they fear they will not do well. This may be because they know they are not prepared. Again, the remedy is simple: Study well. Anxiety can also arise from a bad past experience. If you have done poorly on tests in the past, your self-confidence may be shot so you anticipate doing poorly. That may lead to cramming and memorizing instead of truly learning and may cause you to become excessively nervous during the test, which can cause poor performance. All you need is a couple of good grades on tests to get your confidence back!

If you are a nervous test taker, do not study for about an hour immediately before your exam. Many students who complain of test anxiety are frantically reviewing their notes right up to the moment they are given the test. They have been trying to quickly glance back over everything while racing against the clock. No wonder they are stressed when they begin the test! Remember that your brain needs time to process the information. When you cram information into the "closet," who knows what will fall out when you open the door during the test.

If you have studied well in advance and don't get nervous at exam time, you might want to glance quickly through your notes beforehand, but only if you have time to do so and still allow *at least* a half hour to relax and mentally prepare for your test. The half hour off allows your brain to process the information while you relax. Try getting a light snack so you are alert—a heavy meal could make you drowsy during the test. Walk around to release nervous energy. Listen to music that makes you happy. Sit comfortably, close your eyes, and breathe deeply and slowly while you picture yourself in a relaxing setting—maybe on a tropical beach, curled up on your couch with a good book, or out on a boat fishing. Focus on how relaxed you feel and try to hold that feeling. Now, staying in that mood, concentrate on how well you have studied and keep reminding yourself that

■ I have prepared well for this test.

■ I know this material well, and I answered all questions correctly while studying.

■ I can and *will* do well on this test.

■ I refuse to get nervous over one silly test, especially because I know I am ready.

■ I am ready and relaxed. Let's get it done!

TEST-TAKING TIPS

There are also strategies you can use while taking the test. Let's see what your current strategies are. Complete the survey in **Table 1.6**; then we will discuss specific strategies.

During an exam, be careful—read each question thoroughly before you answer. This is especially true of multiple choice and true/false questions. We know the answer is there, so our eyes tend to get ahead of our brains. We skim the question and jump down to the answers before even trying to mentally answer the question. Slow down and think before moving to the answers. Otherwise you may grab an answer that sounds familiar but is incorrect. If you have trouble keeping your eyes off the answers, cover them with your hand until you finish reading the question and think of the answer on your own.

If you do not know the answer initially, take a deep breath and think of all you do know about the words in the question. Often this is all you need to recall the answer. This is when those concept maps you made will really come through for you.

Use the process of elimination. If you are not sure which answer is correct, can you eliminate any you know are incorrect? Narrow down your choices. Avoid making a guess unless the process of elimination fails you; however, guessing is usually better than leaving a question unanswered, unless you lose points for wrong answers. On short-answer, fill-in-the-blank questions, and essays, always write something.

After you answer a question, read your answer to be sure it says what you want it to; then leave it alone. Once you move on, avoid the temptation to go back and change your answers, even those of which you were unsure. Often we have a gut instinct to write the correct answer;

TABLE 1.6   Self-evaluation of test-taking skills. For each of the following valuable test-taking skills, mark if you do each one always, sometimes, or never. Highlight any that you do not currently use that you think might help you be more successful.

Test-taking Skill	Always	Sometimes	Never
1. While studying my notes and the book, I think of and answer possible test questions.			
2. I use online practice quizzes when they are available.			
3. I avoid last-minute cramming to avoid confusing myself.			
4. I scan the whole test before starting to see how long it is and what type of questions it contains.			
5. I do the questions I am sure of first.			
6. I budget my time during a test so I can complete it.			
7. I answer questions with the highest point values first.			
8. I read all answer options on multiple-choice questions before marking my answer.			
9. I know what key words to look for in a multiple-choice question.			
10. I use the process of elimination during multiple-choice or matching tests.			
11. I know what key words to look for in essay questions.			
12. I look for key words like *always, never,* and *sometimes.*			
13. When I am unsure of an answer, I go with my first answer and fight the urge to change it later.			
14. I try to answer everything even if I am uncertain, instead of leaving some questions blank.			
15. I check my answers before turning in a test. I reread the essay or short answer questions so I'm sure I answered all of the parts.			

perhaps we are recalling it at some subconscious level. But the very act of going back is a conscious reminder of uncertainty, and we more often choose something different only because we doubt ourselves.

When answering multiple-choice or true/false questions, ignore any advice that suggests that a particular answer is more likely to be correct than others. Also, don't worry if you choose the same answer several times in a row, thinking the instructor would not structure a test that way. We can't speak for all instructors, but most don't give much thought to the pattern the answers will make on the answer sheet, so neither should you.

Here are a few more pointers:

- Note the wording on questions. Key words to look for that can change an answer are *always, sometimes, never, most, some, all, none, is,* and *is not.*

- Glance over the exam as soon as you receive it so you know what to expect; then budget your time accordingly.

- Look for questions on the backs of pages so you don't miss them.

- Tackle easy questions first. They may provide hints to the tougher ones and build your confidence.

- Be aware of point values and be sure the questions with the greatest point values are done well. Often essay questions—which usually are worth more points—are at the end, and some students run out of time before reaching them, losing significant points and seriously hurting their grade.

- If you have trouble writing essay answers, recall all you know about the topic, organize in your mind how you would explain it to someone, then write down your thoughts as if you are writing yourself a script on what to say.

- If a question has multiple parts, be sure to answer each part. This is especially true for essays.

- If you are asked for a definition, give a book explanation of what the term or concept means. If you are asked for an example, list an example and explain why it is an example of the concept. If you are asked to explain a concept or term, approach it as if you are trying to teach it to a 6-year-old. Assume the reader has no prior knowledge.

- When asked to compare and contrast two things, be sure you discuss how they are similar and how they are different. Don't just define the two things and assume your instructor will think you know the similarities and differences.

■ Be thorough and specific in your answers. The grader cannot get inside your head to decide if you knew it or not, so your words must literally convey your meaning.

When a test is returned, record your grade. Be sure to review the test to see which questions you missed and why you missed them, and make notes to go back and review that material. Remember—this information may come back to haunt you on a bigger test or on the final exam, and you should know it anyway.

✔ **QUICK CHECK**

How can you slow yourself down when taking a multiple-choice or true/false test? _____

_____

Answer: Cover the answers with your hand while you read the question, and don't look at them until you think of the answer.

Through the Looking Glass: **Individual Accountability**

We hope you have gained insight into the learning process and developed new strategies to improve your success, not just in microbiology, but in all of your classes. One area remains to discuss, though, and that is your responsibility and attitude. When we get frustrated, we often look elsewhere for the cause, even when it may be right on top of our own shoulders. Attitude can transform poor students into honors students, and honors students into dropouts. Many factors can contribute to these changes, but a common thread is always attitude and accountability. Here are three facts you need to firmly implant in your mind:

1. *You*, and nobody else, chose to pursue this academic path.

2. *You*, and nobody else, are responsible for attaining the success you desire.

3. *You*, and nobody else, earn the grades you get.

You must do everything you can to guarantee your success—nobody will do it for you. That means always accepting responsibility for your own effort. No excuses. To stay on track, you must know exactly what you want and always stay focused on where you are going. At times, you may not feel like you can keep up, but instead of quitting or slacking off, you need to refocus on where you are going and why it matters to you. Always set short-term and long-term goals. Write them down and post them where you will see them often. You are responsible for keeping yourself motivated. Learn to visualize your academic success. Think about how your life will be. Dream big. Then go after that dream with all you have.

Final Stretch!

Now that you have finished reading this chapter, it is time to stretch your brain a bit and check how much you learned.

WHAT DID YOU LEARN . . .

PART A: IN THE LEFT-HAND COLUMN THAT FOLLOWS, WRITE YOUR APPROACH BEFORE READING THIS CHAPTER. IN THE RIGHT-HAND COLUMN, LIST ANY CHANGES YOU PLAN TO MAKE TO ENSURE YOUR SUCCESS IN THIS CLASS.

What I have done before this chapter	What I will do to improve
During lectures:	
Note taking:	
Study habits:	
Textbook reading:	
My study place:	
Time management:	
Test taking:	

PART B: LIST THE THREE AREAS IN WHICH YOU THINK YOUR STUDY SKILLS ARE THE WEAKEST AND WAYS IN WHICH YOU PLAN TO IMPROVE THEM.

1.

2.

3.

PART C: WRITE THE CHAPTER TERMS IN A NOTEBOOK AND DEFINE THEM IN YOUR OWN WORDS. GO BACK THROUGH THE CHAPTER TO CHECK YOUR MEANINGS, CORRECTING THEM AS NEEDED. LIST EXAMPLES WHEN APPROPRIATE.

Visual learner
Auditory learner
Tactile learner
SQR3
Active learner
Concept map
Active learning
Critical thinking

2 Basic Math Review

Crunching the Numbers

When you complete this chapter, you should be able to

- Solve math problems involving multiple operations, fractions, decimals, and percents.

- Calculate a mathematical mean.

- Work with exponents, numbers in scientific notation, and ratios.

- Understand the units of the metric system.

- Perform basic measurements in a laboratory setting.

- Read tables, graphs, and charts.

Your Starting Point

Answer the following questions to assess your math skills.

1. $2/3 \times 3/4 =$ _____

2. Express 50% as a fraction: _____
   As a decimal: _____

3. What is the *mean* of 27, 33, and 36? _____

4. In scientific notation, $1000 = 10$ ———

5. 30% of 200 = _____

6. Which is longer, 1 yard or 1 meter? _____

7. In the metric system, the basic unit of weight is the _____.

8. Assume you take 15 quizzes by the end of the semester and get 9 As and 6 Bs. Express this as a ratio. _____

9. Which of these units are used to measure volume:
   a) millimeters   b) milliliters   c) milligrams

10. On a graph, the vertical line is the _____ axis.

How Much Wood Would a Woodchuck Chuck? **Math in Science**

You probably remember doing story problems when learning math in your younger years. Those problems were to help you see how math can be used. Many students are surprised to learn that they have to use math in biology. But you must remember that science—all science—deals with that which is testable. A scientific test, as you know, is called an experiment. Results collected from experiments are called **data** and, more often than not, the data are numbers. When you try to make sense of the data, you are working with numbers, and that means math.

Answers: 1. 1/2 2. 1/2, 0.5 3. 32 4. 10^3 5. 60 6. 1 meter 7. gram 8. 9:6 = 3:2 9. b 10. y

You will likely do some experiments in lab and then analyze the data. These experiments will require you to deliver certain volumes of a solution to a test tube or measure the length of a bacterium. You may be required to calculate the total magnification of an object seen in your microscope. You might also have to count the number of colonies growing in a Petri plate and use this to determine the concentration of bacteria in the source from which they came. Your textbook will contain graphs and charts, and you'll need to be able to understand what they mean. In addition, you may want to figure out your grade at a certain point in the semester.

Many students entering this class may only need a brief reminder of what they learned before, whereas others may need to learn it again. Regardless of your math history, a quick refresher will help you better understand the numbers.

From the Beginning: **Basic Math Operations**

You might need to do some complicated computations in class, so it is good to remember the basic rules. Let's zip through multiplication and division for a quick refresher.

MULTIPLICATION   Multiplication problems can be done in any order: $3 \times 4 \times 2 = 24$, or $2 \times 3 \times 4 = 24$. The answer to the equation is called the **product**. Recall that multiplying any number by 1 does not change the number, while multiplying any number or numbers by 0 gives you 0.

Multiplication problems are sometimes represented with **exponents**. Consider 2^4, which is read as "two to the fourth power," and 10^3, which is read as "ten to the third power." These examples are really just a shorthand way of expressing these multiplication problems:

$$2^4 = 2 \times 2 \times 2 \times 2 = 16 \qquad 10^3 = 10 \times 10 \times 10 = 1000$$

DIVISION   Like subtraction, division problems must be done from left to right. In division, the number being divided is called the **dividend**, the number by which it is divided is the **divisor**, and the total is the **quotient**. In our first example, 10 is the dividend, 5 is the divisor, and 2 is the quotient.

$$10 \div 5 = 2,$$
$$\text{while } 5 \div 10 = \frac{1}{2}$$

MULTIPLE OPERATIONS   Let's try some more complicated problems that involve more than one mathematical operation. Parentheses or brackets are often used in equations when there are multiple operations. Think of them as directors telling you how to proceed. Remember, always do operations within these structures first. Let's see why this matters.

$$8 - (2 \times 3) = 8 - (6) = 2,$$
$$\text{but } 8 - 2 \times 3, \text{ done in that order,} = 6 \times 3 = 18$$
$$\text{OOPS!}$$

If a number appears immediately to the left of a parenthesis or bracket, multiply by that number even though there is no multiplication sign:

$$3(6 - 2) = 3 \times (6 - 2) = 3 \times 4 = 12$$

Always approach an equation by first doing any operations within parentheses. ■

Here are some simple rules to help ensure that you perform mathematical operations in the correct order:

1. First, do all operations inside the parentheses or brackets.

2. Next, multiply out any exponents.

3. Then do all multiplication and division equations, moving from left to right.

4. Finally, do all addition and subtraction problems, again from left to right.

TIME TO TRY

Try the following problems.

1. $4 \times (9 - 6) + 10 =$ _____

2. $3^3 \div 9 - 4 + 5 =$ _____

3. $6^2 - 2(5 - 2) + 4 - 2 =$ _____

Let's see how you did.

Problem #1: The correct answer is 22. First, do what is in the parentheses (rule 1): $(9 - 6) = 3$, so the problem becomes $4 \times (3) + 10$. (Note, now we can drop the parentheses around the 3.) Next, do the multiplication (rule 3): $4 \times 3 = 12$, so the problem becomes $12 + 10$. Finally, do the addition (rule 4): $12 + 10 = 22$.

Problem #2: The correct answer is 4. There are no parentheses, so you start with the exponent (rule 2): $3^3 = 3 \times 3 \times 3 = 27$, and the problem becomes $27 \div 9 - 4 + 5$. Next, do the division (rule 3): $27 \div 9 = 3$, so the problem becomes $3 - 4 + 5$. Finally, do the addition and subtraction from left to right (rule 4), and you get $3 - 4 = -1$; then $-1 + 5 = 4$.

Problem #3: The correct answer is 32. Start in the parentheses (rule 1): $(5 - 2) = 3$, so the problem becomes $6^2 - 2(3) + 4 - 2$. Next, take care of the exponent (rule 2): $6^2 = 6 \times 6 = 36$, so the problem becomes $36 - 2(3) + 4 - 2$. Now, do the multiplication and division from left to right (rule 3): $2(3) = 2 \times 3 = 6$ and $4 - 2 = 2$, so the problem becomes $36 - 6 + 2$. Finally, do the addition and subtraction from left to right (rule 4): $36 - 6 = 30$, then $30 + 2 = 32$.

As you see, some mathematical equations can be long and somewhat complicated, but if you keep the basic rules in mind and tackle them step by step, they become quite manageable.

What Do You **Mean** You Are Normal?

Do you know what "normal" body temperature is? Sure you do—98.6°F. You have likely known that since you were a small child. But what is *your* normal temperature? What if yours is rarely 98.6°? "Normal" blood pressure is 120/80. What if yours runs lower than that? If you think this makes you abnormal, reconsider what the term "normal" really indicates.

In biology, the term ***normal*** means ***average***. And in math, another term for average is ***mean***. We refer to many normal values—temperature, blood pressure, pulse, respiratory rate . . . the list goes on and on. When you see these, realize that they are average values and an individual person may have a different normal value—what is normal for him or her may not be average for the whole population. To better understand this idea of normal, you need to know how to calculate the average, or mean, value.

Let's say you have earned the following scores on your first three quizzes in microbiology:

Quiz #1: 74%   Quiz #2: 84%   Quiz #3: 88%

Excellent! You're improving!

What is your average score on the microbiology quizzes? To find the mean of a group of numbers, simply add them all together and then divide the total by how many numbers you added. For your data, you would add the three scores and divide by 3:

74% + 84% + 88% = 246%   246% ÷ 3 = 82%

Did you notice that the mean is not one of the original numbers? It does not have to be. It is the average of all three numbers.

Here is a tip to help you with means and with most math problems— learn to predict your results. The mean of a group of numbers will be somewhere between the highest and the lowest of the numbers you are averaging. If your value does not fall in that range, check to see if you made an error. Common errors are missing a number during the addition or dividing by the wrong number. If you estimate your result first, you can more easily recognize errors if they occur.

TIME TO TRY

Five teams of microbiology students performed a test to learn how many bacterial colonies were produced from 1 mL of a water sample. The teams counted 53, 95, 48, 62, and 77 colonies.

1. Without actually doing any calculation, predict the mean.

_____

2. Calculate the mean of the colony counts. _____

_____

3. What does this mean *mean*? _____

_____

Because 48 is the lowest number and 95 is the highest, the mean would be a number that falls between these two numbers. If you did this correctly, you should have gotten a mean of 67 colonies, even though that was not one of the original counts listed. So, as stated earlier, the "normal" value is a mean, and individual sample colony counts will vary around that mean.

Meet My Better Half: **Fractions, Decimals, and Percents**

Working with whole numbers is rather easy and is second nature to most of us. Some of us may need to brush up on fractions, however. Related to fractions are two other ways of expressing values: decimals and percents. In microbiology, all three of these will be used. For example, the micrometer (μm), a common unit for measuring the size of bacteria, is a tiny fraction of the more familiar millimeter—a micrometer is 1/1000 of a millimeter, to be exact. As you just saw, normal body temperature is reported as a decimal (98.6). Finally, about 80% of the cytoplasm of a cell is water. You will discover that many biological values are reported in one of these formats, so you want to be comfortable with their use.

FRACTIONS

Fractions are written as *a/b*, in which *a* and *b* are both whole numbers, and *b* is not 0. The first (top) number is called the **numerator**, and the one on the bottom is the **denominator**. A fraction represents parts of some whole group (**Figure 2.1**). For example, 3/4 represents 3 equal parts out of 4 equal parts, where the 4 equal parts make up the whole (Figure 2.1a). Whole numbers can be represented as fractions as well. The whole number simply becomes the numerator, and the denominator is 1, so 3 = 3/1.

✔ **QUICK CHECK**

For 5/8, what is the numerator? _____ The denominator? _____
Express 6 as a fraction: _____

Answers: 5 is the numerator, 8 is the denominator, and 6 as a fraction is 6/1.

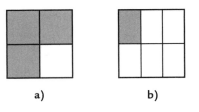

a)   **b)**

FIGURE 2.1   Fractions represent some part of a whole. In each of these squares, the shaded area represents the fraction, as follows: **a)** This square is divided into four equal parts, and 3 of the four are shaded = 3/4. **b)** This square is divided into six equal parts, and 1 of the 6 is shaded = 1/6.

REDUCING FRACTIONS   **Equivalent fractions** have the same value even though they appear to be different. Consider the following fractions:

$$\frac{1}{3} \qquad \frac{2}{6} \qquad \frac{4}{12} \qquad \frac{7}{21}$$

All of these numbers have the same value: 1/3. To see this, you need to **reduce** the other fractions. This is done by finding the **greatest common factor (GCF)** for each fraction. The greatest common factor is the largest whole number that can be divided into both the numerator and the denominator. Consider 2/6. Both the numerator (2) and the denominator (6) are divisible by 2, which is the greatest common factor. If you do the division, you see that 2 ÷ 2 = 1 and 6 ÷ 2 = 3, so 2/6 becomes 1/3. So, 1/3 is the reduced fraction.

TIME TO TRY

Look at the other fractions we listed: 4/12 and 7/21.

What is the greatest common factor for 4/12? _____

Divide the numerator by that factor: _____
Divide the denominator by that factor: _____

What is the reduced fraction? _____

What is the greatest common factor for 7/21? _____

Divide the numerator by that factor: _____

Divide the denominator by that factor: _____

What is the reduced fraction? _____

How did you do with Time to Try? You should have found that the greatest common factor for 4/12 is 4, so 4 ÷ 4 = 1 and 12 ÷ 4 = 3. Thus, the fraction 4/12 reduces to 1/3. Similarly, 7/21 has a greatest common factor of 7, and 7 ÷ 7 = 1, and 21 ÷ 7 = 3 so, again, 7/21 reduces to 1/3. After using a number to reduce the fraction, check your result to see if it is in its simplest form or if it can be reduced further.

✔ **QUICK CHECK**

What is the most reduced form of each of the following fractions: 60/90, 25/100, and 18/54?

Answer: 60/90 has a GCF of 30 and reduces to 2/3; 25/100 has a GCF of 25 and reduces to 1/4; 18/54 has a GCF of 18 and reduces to 1/3.

MULTIPLYING AND DIVIDING FRACTIONS   When doing mathematical operations with fractions, the rules are different for multiplication and division than they are for addition and subtraction. For multiplication, you simply multiply the numerators in one step, then multiply the denominators and then reduce. Consider 2/3 × 3/4. The numerators are 2 and 3. Multiply them to get 6, and that goes on top. Next, multiply the two denominators, 3 × 4, to get 12. So the product is 6/12, which reduces to 1/2:

$$\frac{2}{3} \times \frac{3}{4} = \frac{6}{12} = \frac{1}{2}$$

Let's try another: 1/3 × 2/5 × 3/4 = _____

First, multiply all the numerators (1 × 2 × 3 = 6); then multiply all the denominators (3 × 5 × 4 = 60), and you get 6/60, which reduces to 1/10.

To multiply fractions, first multiply all the numerators and then multiply all the denominators. Reduce the result as needed. ▪

Dividing fractions may seem difficult at first, but a simple trick actually makes it easy! These problems may be written two different ways:

$$\frac{\frac{4}{5}}{\frac{2}{3}} \quad \text{or} \quad \frac{4}{5} \div \frac{2}{3}$$

Solving them is easy. First, invert (flip) the second fraction, which is the divisor: 2/3 becomes 3/2. Then you simply multiply the two fractions:

$$\frac{4}{5} \div \frac{2}{3} = \frac{4}{5} \times \frac{3}{2} = \frac{12}{10}$$

Now, 12/10 can be reduced to 6/5. Here the numerator is larger than the denominator, so express it as a mixed number: 1-1/5. To do this, first reduce the fraction: 12/10 = 6/5. Then realize that 6/5 = 5/5 + 1/5. Considering 5/5 equals 1, the mixed number would be 1-1/5 (read as 1 and 1/5).

 To divide one fraction by another, first invert the second fraction to turn it into a multiplication problem. Next, multiply the numerators and then multiply the denominators. Finally, reduce the result. ▪

ADDING AND SUBTRACTING FRACTIONS   To add or subtract fractions, they must first be in the same format. You might think you can just add the numerators and denominators, but that won't work. By that method, 1/2 + 1/4 would equal 2/6, which reduces to 1/3. But that is smaller than 1/2, one of the numbers we added! This doesn't make sense. (See why it helps to predict your results?) Instead, you must first put the fractions into common terms. They must have the same denominator, called a **common denominator**.

To get the common denominator, you need to know the **least common multiple (LCM)**. This is the smallest number that can be divided by both numerators and both denominators. In our example of 1/2 + 1/4, 4 is the least common multiple, so we want both fractions to have 4 as their denominator. To convert 1/2 into fourths, we multiply it by 2/2 (=1). Recall that any number multiplied by 1 does not change. Thus, 1/2 × 2/2 becomes 2/4. Once the fractions have a common denominator, we simply add the numerators only:

$$\frac{2}{4} + \frac{1}{4} = \frac{3}{4}$$

Figure 2.2 illustrates this for you.

Subtracting fractions also requires a common denominator. Once you have the denominator, simply subtract the second numerator from the first numerator. Let's try one: 1/3 − 1/4. The smallest common denominator for these two fractions is 12, so they both need to be converted, as follows:

$$\frac{1}{3} \times \frac{4}{4} = \frac{4}{12}$$
$$\frac{1}{4} \times \frac{3}{3} = \frac{3}{12}$$

Now line up your fractions in the correct order from left to right and then subtract the second numerator from the first:

$$\frac{4}{12} - \frac{3}{12} = \frac{1}{12}$$

To add or subtract fractions, use a common denominator to put the fractions in a common form and then add or subtract the numerators only. Remember to always subtract from left to right. ∎

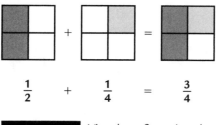

$$\frac{1}{2} \quad + \quad \frac{1}{4} \quad = \quad \frac{3}{4}$$

FIGURE 2.2   Visual confirmation that adding 1/2 + 1/4 = 3/4.

✔ **QUICK CHECK**

Solve these problems:
1. 3/5 + 1/4 + 1/10 = _____
2. 9/16 − 3/8 = _____

Answers: 1. The smallest common denominator is 20, so 3/5 × 4/4 = 12/20, 1/4 × 5/5 = 5/20, and 1/10 × 2/2 = 2/20. Then add the numerators: 12 + 5 + 2 = 19/20. **2.** The smallest common denominator is 16, so 3/8 × 2/2 = 6/16. Then 9/16 − 6/16 = 3/16.

DECIMALS

Decimals are common in microbiology. All decimals are based on 10 in a specific way—each place in the number represents a multiple of 10. The value *increases* 10 times for each space that you move to the left of the decimal, and it *decreases* 10 times for each space to the right.

Let's consider this number: 12345.6789. You know how to read the numerals to the left of the decimal. They make 12,345. Moving from the decimal to the left, you can see how the spaces represent, in order, 1s, 10s, 100s, 1000s, and 10,000s (**Figure 2.3**). Our number represents 10,000 + 2000 + 300 + 40 + 5. The spaces to the right of the decimal represent fractions: tenths, hundredths, thousandths, and so on. Thus in our number, as you go to the right of the decimal, the numerals represent the fractions 6/10, 7/100, 8/1000, and 9/10,000.

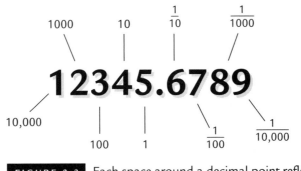

FIGURE 2.3 Each space around a decimal point reflects a change by a factor of 10.

CONVERTING DECIMALS   As with fractions, decimals allow you to express a number more precisely than you can with whole numbers. In fact, you can think of fractions as division problems that give you a decimal value, so fractions can be converted to decimals. For example, 1/2 is 1 ÷ 2, and that equals the decimal 0.5. Note the 0 here—it is used as a place holder so that you know where the decimal belongs. If you use a calculator to find 1 ÷ 3, the answer will be 0.333333 . . . . This is known as a **repeating decimal**. When you have a repeating decimal, one option is to round off. If the last number is less than 5, round down; if it is 5 or higher, round up. For example, 0.33333 . . . would round down to 0.33, and 0.66666 . . . would round up to 0.67.

Decimals can also be converted into fractions. The value 0.25 represents 2/10 + 5/100. When you add them, remember to first get a common denominator. You get 20/100 + 5/100 = 25/100, and that reduces to 1/4.

ADDING AND SUBTRACTING DECIMALS   Adding and subtracting decimals is easy as long as you line them up correctly. Consider this example: 1.287 + 24.32. First, be sure the two numbers have the same number of spaces after the decimal. The first number has three places, but the second only has two. But, you can always add zeros to the end of a decimal number: 24.32 = 24.320. Note that the last 0 means there are 0/1000, which is correct and does not change the value. Finally, it is easiest to add these numbers if you line them up vertically, always being sure the decimal points are aligned:

$$
\begin{array}{r}
1.287 \\
+\ 24.320 \\
\hline
=\ 25.607
\end{array}
$$

Subtraction is done the same way.

PICTURE THIS

Assume that you have worked three extra jobs for pocket cash this week. From them, you earned $33.70, $45.28, and $21.02. How much extra money did you earn? _____

With this money you buy a pizza for $8.99, soft drinks for $1.49, gas for $20.00, and a new CD for $19.95. How much money do you have left? _____

Congratulations, you just added and subtracted decimals, as you do on a regular basis in daily life! You should see that you earned $100 and have $49.57 left.

When adding or subtracting decimals, always align the decimal point in the two numbers before doing the operation. ■

MULTIPLYING AND DIVIDING DECIMALS   Multiplication and division of decimals is a bit trickier because you must keep track of how many decimal places (digits to the right of the decimal point) you should have at the end. Let's try an easy one: 0.5×0.3. First, multiply the numbers as if they are whole numbers: $5 \times 3 = 15$. Now, add the number of decimal places you started with. Both numbers you multiplied originally had one decimal place, so that adds up to two. Realize that your answer of 15 is 15.0, so you know where the decimal begins. Now you have to move the decimal. The numbers you started with had a total of two digits after the decimal, so you must end up with two decimal places. You will have to move the decimal point left by two places, giving you 0.15. Here is a way to double-check that. If the original numbers were fractions, they would be 3/10 and 5/10. Recall how to multiply fractions—you multiply the numerators and then multiply the denominators:

$$\frac{3}{10} \times \frac{5}{10} = \frac{15}{100} = 0.15$$

What if the numbers had been 0.03×0.5? Although you still get 15, now you need to move the decimal three places to the left, but there are only two. You simply add zeros to the left until you have the correct number of decimal places, in this case giving you 0.015.

Division with decimals is just like ordinary division, except we keep going until we either finish or reach a predetermined stopping point.

You usually do not go beyond the number of decimal points your original numbers contain, so if they had a total of two, you would likely stop at two and round off beyond that. Let's look at an example: 2.8 ÷ 7. Put the decimal point in the answer line exactly above its position in the dividend (2.8); then simply do the division:

$$
\begin{array}{r}
0.4 \\
7\,\overline{)\,2.8} \\
-2\,8 \\
\hline
0
\end{array}
$$

How do you divide when both numbers are decimals? All you have to do is move the decimal of the divisor until you have a whole number, and then move the decimal of the dividend by the same number of spaces in the same direction. If you try 1.68 (dividend) ÷ 0.3 (divisor), you move the decimal in 0.3 one spot to the right to get 3. Then you must also move the decimal in 1.68 one spot to the right, getting 16.8, so the problem becomes 16.8 ÷ 3.

$$
\begin{array}{r}
5.6 \\
0.3\,\overline{)\,1.68} \\
-15 \\
\hline
18 \\
-18 \\
\hline
0
\end{array}
$$

You can always multiply back to double-check your result:

$$0.3 \times 5.6 = 1.68$$

Moving the decimal may seem confusing, but there are some easy shortcuts to remember.

■ Moving the decimal to the *right* one space is the same as multiplying by 10; moving two spaces multiplies your number by 100, and so on, so the numbers get *bigger*.

■ Moving the decimal to the *left* means you are dividing by 10 for each space moved, and the number always gets *smaller*.

■ Always ask yourself if your answer seems reasonable.

Moving the decimal to the right is the same as multiplying by 10 for each space moved. Moving it to the left is the same as dividing by 10 for each space moved. ■

✔ **QUICK CHECK**

Solve these problems:

1. 1.27 + 3.6 = _____
2. 14.87 − 3.2 = _____
3. 2.4 × 1.2 = _____
4. 8.4 ÷ 0.2 = _____

Answers: 1. 1.27 + 3.60 = 4.87 2. 14.87 − 3.20 = 11.67 3. 2.4 × 1.2 = 2.88
4. Move the decimal in the divisor and dividend both one space to the right, so it becomes 84 ÷ 2 = 42.

PERCENTS

As you learn microbiology, you will encounter many values that are stated as percents. Percents are based on 100, where 100% is the total. For that reason, when working with percents, always be sure they add up to 100% and no more than that.

Percents are easy to work with. They are essentially fractions expressed as hundredths. For example, 25% is the same as 25/100, which can be further reduced to 1/4. And because fractions can be expressed as decimals, so can percents. You simply put the decimal two places to the left of the percent. So, 25% becomes 0.25 and 7% becomes 0.07. Percents, decimals, and fractions are all interchangeable, but when doing math operations, percents should be converted into either decimals or fractions. You cannot do math operations with mixed expressions—they must all be whole numbers, or fractions, or decimals. **Table 2.1** explains the relationship between these expressions.

Percent	Decimal	Fraction in hundredths	Reduced fraction
10%	0.10	10/100	1/10
25%	0.25	25/100	1/4
40%	0.40	40/100	2/5
50%	0.50	50/100	1/2
75%	0.75	75/100	3/4
100%	1.00	100/100	1

TABLE 2.1 The relationship between percents, decimals, and fractions.

TIME TO TRY

Can you supply the missing information in this table?

Percent	Decimal	Fraction
36%	_____	_____
_____	0.42	_____
_____	_____	80/100 = 4/5

Answers: 36% = 0.36 = 36/100 = 9/25; 42% = 0.42 = 42/100 = 21/50; 80% = 0.8 = 80/100 = 4/5.

Can You Feel the Power? **Understanding Exponents**

We briefly discussed exponents earlier when we stated that a number written with an exponent is basically a multiplication problem: $2^3 = 2 \times 2 \times 2 = 8$. In science, very large and very small numbers are often written in a special format that uses exponents based on powers of 10. This format is called **scientific notation**. Let's consider the number 200 to see how this format is used. To write a number in scientific notation, first place the decimal immediately after the first digit, and then drop the zeroes. This number is called the **coefficient**. In this example, the coefficient is 2. Next, count how many spaces you moved the decimal—two places to the left. Each of those spaces represents a power of 10, so two places means $10 \times 10 = 100$. In scientific notation, we would write 200 as 2×10^2. As you can see, this is $2 \times 10 \times 10 = 200$. Remember, the first number in scientific notation must be greater than 1 but less than 10. If a number is less than 1, the exponent is a negative power of 10. For example, 0.0004 would be 4×10^{-4} because the decimal was moved four spaces to the right. **Table 2.2** lists some common exponents.

TABLE 2.2   The values of some common exponents used in scientific notation.

Exponent	Value	Term
10^9	1,000,000,000	billions
10^6	1,000,000	millions
10^3	1000	thousands
10^2	100	hundreds
10^1	10	tens
10^0	1	ones
10^{-1}	$\dfrac{1}{10}$	tenths
10^{-2}	$\dfrac{1}{100}$	hundredths
10^{-3}	$\dfrac{1}{1000}$	thousandths
10^{-6}	$\dfrac{1}{1,000,000}$	millionths
10^{-9}	$\dfrac{1}{1,000,000,000}$	billionths

TIME TO TRY

Express the two numbers below in scientific notation.

24,000,000 = _____

0.003 = _____

If you did this correctly, you got 2.4×10^7 and 3×10^{-3}.

My Cell's Bigger than Your Cell! Ratios

- Carbohydrates contain hydrogen and oxygen in a 2:1 ratio.

- In the United States, the ratio of males to females at birth is about 105:100.

- The ratio of males to females declines steadily until, after age 85, it is only 40.7:100.

HUH? Welcome to the comparatively interesting realm of ratios. A **ratio** expresses a relationship between two or more numbers—it is a way to compare them. Ratios can be expressed using a colon between the numbers (as above), as a fraction, or by using the word "to." In the first example above, 2:1 means there are twice as many hydrogen atoms in carbohydrates as there are oxygen atoms.

Ratios are used for comparison, and they can also be expressed as fractions. Let's say the ratio of men to women in your microbiology class is 1:2. We're not saying that half of the class consists of men, we are saying there are half as many men as women. To convert this ratio to a fraction, we use the first number as the numerator and the sum of both numbers as the denominator. Thus, 1/3 of the class consists of men.

If ratios can be expressed as fractions, they can also be expressed as decimals and percents. Because they can be written as fractions, they can also be reduced like fractions. For example, a ratio of 4:6 is the same as 4/10, which is the same as 2/5. When working with ratios, it is critical to write them in the correct order. If a microbiology class has 10 males and 20 females, the ratio of males to females is 10:20. If we write it as 20:10, it means there are twice as many males as females, which is not true.

TIME TO TRY

Empty your pocket or purse of change. Separate the coins by denomination. Count all of the coins in each category. Now express those numbers in a ratio: _____

Why is it important to indicate the order in which you are listing the coins? _____

_____

If you did this correctly, you should have indicated the order of the coins because without that reference, we have no idea what number corresponds with which coin. Perhaps you had 5 pennies, 4 nickels, 3 dimes, and 2 quarters. If you wrote your ratio in that order, it would be 5:4:3:2.

We also use ratios to discuss quantities in a certain amount. For example, there are about 280 million hemoglobin molecules in each red blood cell. That can be expressed as 280 million/cell, which is read as 280 million per cell; it is really saying the ratio is 280 million to 1.

Who Ever Heard of a Centimeter Worm? **The Metric System**

In the United States, we all grew up learning there are 12 inches to a foot, 3 feet to a yard, and 100 yards to a football field, and we measure driving distances in miles, which contain 5280 feet. In the kitchen, we use cups, tablespoons, teaspoons, pints, quarts, gallons, ounces, and pounds. There are so many units in our system it is amazing we can keep them straight. But of course there is a simpler way to measure.

It is called the **metric system**, or **International System of Units** (abbreviated SI from the French Le Système International d'Unités). It is universally used in science and all but three countries, including the United States. You have undoubtedly had brushes with learning the metric system, and you may have found it difficult. The problem is not with the metric system, which is delightfully simple. Instead, the problem is with our complicated U.S. (also called English) system and the need to convert between the two systems. This requires—you guessed it—math.

WHY SHOULD I CARE?

Science uses metric measurement almost exclusively, so you will need a basic understanding of metric units for all your future course work. In addition, almost everyone on our planet—except those in the United States—uses the metric system.

The metric system is amazingly simple because it is all based on the number 10, and decimals are easier to use than the fractions used in the United States. For our purposes, we'll learn four main units used in science: those that deal with length or distance, mass, volume, and temperature. Each of these has a standard or base unit:

- The basic unit of length (or distance) is the **meter (m)**.

- The basic unit of mass is, technically, the kilogram (kg), but many sources use the **gram (g)** as the base unit instead.

- The basic unit of volume is the **liter (L)**.

- The basic unit of temperature is the **degree Celsius (°C)**.

These are the base units, but more convenient units are derived from these. For example, a meter is just a bit longer than 3 feet (39.34 inches), so it is not a convenient unit for measuring the size of a cell. Smaller units of the meter, based on the powers of 10, are used instead. These units are named by adding the appropriate prefix to the term *meter* (**Table 2.3**). *Centi-* means 1/100, and there are 2.54 centimeters (cm) in an inch, so centimeters work well for measuring tapeworms. Cells are microscopic, so they are best measured in even smaller units, such as micrometers—one micrometer = 1 millionth of a meter. Again, all metric units are based on 10. Think about that—you first learn to count from 1 to 10 and then you can count by tens to 100, then by hundreds to 1000, and so on. It is an easy system.

| TABLE 2.3 | Metric system prefixes. |

Prefix	Abbreviation	Decimal equivalent (multiple)	Exponential equivalent (scientific notation)
Pico-	p	0.000000000001	10^{-12}
Nano-	n	0.000000001	10^{-9}
Micro-	μ	0.000001	10^{-6}
Milli-	m	0.001	10^{-3}
Centi-	c	0.01	10^{-2}
Deci-	d	0.1	10^{-1}
no prefix		1.0	10^{0}
Deka-	D	10.	10^{1}
Hecto-	H	100.	10^{2}
Kilo-	k	1000.	10^{3}
Mega-	M	1,000,000.	10^{6}
Giga-	G	1,000,000,000.	10^{9}

Table 2.3 provides many of the prefixes and their base-10 equivalents. In microbiology, you will use some units more often than others. For length or distance, you will mostly work in millimeters (1/1000 m), micrometers, or nanometers. For mass, which is the actual physical amount of something, you will most often refer to grams or milligrams (1/1000 g). For volume, which refers to the amount of space something occupies, the most common units will be liters, milliliters, and microliters.

The Celsius temperature scale does not use prefixes. Instead, it has a single unit: degrees Celsius. However, the scale is different than the Fahrenheit scale with which you are familiar. Water freezes at 0°C and boils at 100°C (**Figure 2.4**).

At what Fahrenheit temperature does water freeze? _____ Boil? _____ In the Celsius scale, normal body temperature (98.6°F) is 37°C. To convert between the two temperature scales, there are two specific equations—one to convert from degrees Celsius to degrees Fahrenheit, and another to convert in the opposite direction. Each of these equations is listed below, first in its original form, which includes a fraction, and then with the fraction converted to a decimal. You will likely use a calculator to do any conversions, and it will be easier to multiply using the decimal.

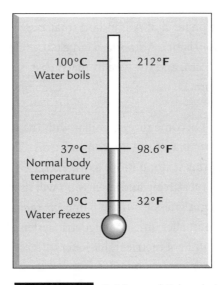

FIGURE 2.4   Celsius and Fahrenheit temperature comparison.

$$°\text{Celsius} = (°\text{Fahrenheit} - 32) \times \frac{5}{9}$$

$$= (°\text{Fahrenheit} - 32) \times 0.556$$

$$°\text{Fahrenheit} = °\text{Celsius} \times \frac{9}{5} + 32$$

$$= °\text{Celsius} \times 1.8 + 32$$

Let's try one of each type of conversion. To convert 37°C to degrees Fahrenheit, we use the second equation:

$$°\text{F} = 37°\text{C} \times 1.8 + 32$$
$$= 66.6 + 32$$
$$= 98.6°\text{F}$$

Now let's convert 68°F to degrees Celsius. We use the first equation as follows:

$$°\text{C} = (68°\text{F} - 32) \times 0.556$$
$$= 36 \times 0.556$$
$$= 20.0°\text{C}$$

Most of the work you will do, though, will be converting units for length, mass, and volume, so let's move on.

Know the paper clip! A standard small paper clip has a mass of about 1 g (it's very light). A standard large paper clip has a side-to-side width of about 1 cm, and the wire from which it is made has a diameter of about 1 mm. ■

You will become more familiar with the metric system as you use it. Rarely, you may need to convert from U.S. units to metric units, although this is done more as an exercise than out of need—in class almost all measuring and discussion will use metric units.

You might guess from the name *microbiology* that you will be using mostly the smaller units in the metric system. Let's try some conversions between different metric units. Refer to Table 2.3 if needed. Let's convert 6 mm into micrometers.

$$6 \text{ mm} = \underline{\hspace{2cm}} \text{ μm}$$

A micrometer is 1/1000 of a millimeter, so there are 1000 micrometers per millimeter. Thus:

$$6 \text{ mm} \times 1000 \text{ μm/mm} = 6000 \text{ μm}$$

Now we will convert 13 μm into millimeters, but let's try another method. All we really have to do to convert between different metric units is move the decimal, but by how many spaces and in which direction? How many spaces you move is determined by the difference in the power of 10. We know that a micrometer is a thousandth of a millimeter:

$$1 \text{ μm} = 10^{-3} \text{ mm}$$

So, we moved the decimal in our number (13), by three spots. But in which direction? When converting from smaller to larger units, the decimal moves left. When converting from larger to smaller units, the decimal moves right. Back to our example: Converting 13 μm to mm gives us 0.013 mm.

When converting within metric units:

1. Put the units in scientific notation and subtract the smaller exponent from the larger one. The difference is how many spaces the decimal will move in your coefficient.

2. If you are converting from small units to larger ones, the number gets smaller, so the decimal moves to the left. If you are converting from larger units to smaller ones, the number gets bigger, so the decimal moves to the right. ▨

TIME TO TRY

Now that you see the simple secret to this process, complete the following conversions:

5 cm = _____ mm   8 mL = _____ μL   6 μm = _____ mm

If you did these conversions correctly, you should see that 5 cm = 50 mm; 8 mL = 8000 μL; 6 μm = 0.006 mm. See, the metric system is easy!

How Do You Measure Up? **Basic Measurement**

Now that you understand the basic units of measurement, you need to know how to measure. A common error in scientific experimentation is called human error, which includes math mistakes (which you won't make now!) and something as simple as not measuring correctly. When you're baking brownies, adding extra sugar and chocolate may be a good thing, but that won't work in science. Measurements must be done precisely and with appropriate equipment.

MEASURING LENGTH

Length is usually measured with a metric ruler (see **Figure 2.5a**). There might be two scales—inches and metric. We're going to use only the metric scale. Notice that there are 10 mm in 1 cm. When measuring with one of these rulers, make sure you are using the correct scale: mm or cm.

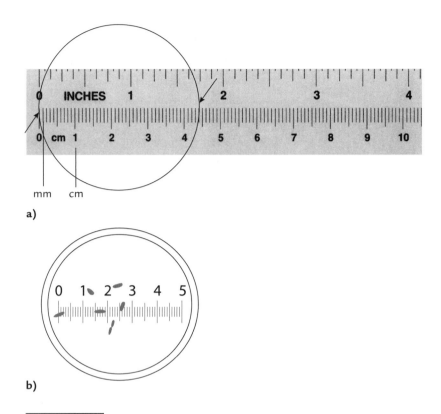

a)

b)

FIGURE 2.5 **a)** This ruler shows the comparison of the metric and inch scales. Millimeters and centimeters are labeled. When measuring an object such as the circle diameter, position the ruler carefully with the zero at the edge of the object. Read the scale at the other edge. **b)** An ocular micrometer might look like this when viewed through the microscope while looking at bacteria. Note that the numbers do not have units. This micrometer needs to be calibrated in order to determine the value of each mark on the scale.

TIME TO TRY

How many millimeters are there in 1 cm? _____
What is the diameter of the circle in Fig. 2.5a in centimeters? _____ cm
What is the diameter of that circle in mm? _____ mm

There are 10 mm in 1 cm. The diameter of the circle is 4.4 cm or 44 mm. In the case of millimeters, you could have counted the number of small marks or used this calculation:

$$4.4 \text{ cm} \times 1000 \text{ mm/cm} = 4.4 \times 100 \text{ mm} = 4,400 \text{ mm}$$

In microbiology, you will be looking at bacteria through the microscope. They are too small to see with the unaided eye. Do you think the ruler in Figure 2.5a would be useful in measuring the size of these bacteria? No! Therefore, some microscopes come with a built-in "ruler" called an ocular micrometer (see **Figure 2.5b**). When you look through the microscope, you see this micrometer and can then align the bacterium with the scale to measure it. You'll notice that the micrometer doesn't have any units (millimeters, centimeters, or others). The length that each mark on the micrometer represents must be determined by calibration and recorded. Then, you simply count the number of marks and multiply by that factor. Let's try an example. If the length of a bacterium is equal to 5 small marks, and each small mark is equivalent to 0.4 µm, then:

$$5 \text{ marks} \times 0.4 \text{ µm/mark} = 5 \times 0.4 \text{ µm} = 2.0 \text{ µm}$$

When measuring length, be sure you use the appropriate scale. ■

MEASURING VOLUME

Volume refers to the amount of space a substance occupies. In microbiology labs, you will most often measure liquid volumes by using pipettes. Always read the scale, usually in milliliters, at eye level for accuracy. You should also know how to read the meniscus (**Figure 2.6**)—this is critical in a pipette. Liquid in a pipette tends to climb slightly up the sides, so the center is lower than the edges where the liquid contacts the container. This dip is called the **meniscus**. When reading the scale, always read it at the low point of the meniscus for the best accuracy.

TIME TO TRY

Find the narrowest clear container you can and fill it halfway with water. Look at it at eye level. Use a ruler on the outside of the container to measure the highest point of the water. _____ Now measure the lowest point. _____ The dip that you see is the meniscus. Whenever you measure a liquid volume, always measure at the lowest point of the meniscus.

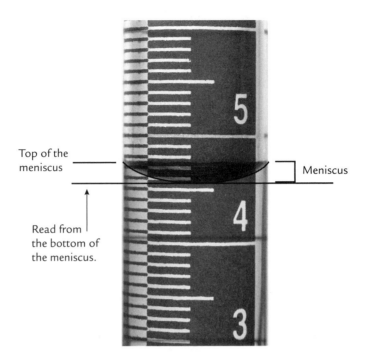

FIGURE 2.6   Read measurements of liquid volumes from the bottom of the meniscus. By doing so, you see that this pipette contains 4.55 mL of liquid, not the 4.75 indicated by the top of the meniscus.

You Ought to Be in Pictures: **Tables, Graphs, and Charts**

We explored how to get numbers by measuring and how to work with them. Now we will see how these numbers and other information, collectively called data, can be depicted.

TABLES

We have already used tables in this book, and you should be familiar with them, so let's quickly review the basics. Tables come in many forms and are a convenient way to present information so it is easy to read and compare. We will use **Table 2.4** as a reference.

When viewing a table, start with the table title, in this case "Estimated deaths of persons with acquired immunodeficiency syndrome (AIDS), by year of death and U.S. region, 2002–2006." The title usually tells you what the table contains. Next, realize that tables are carefully arranged

TABLE 2.4   Estimated deaths of persons with acquired immunodeficiency syndrome (AIDS), by year of death and U.S. region, 2002–2006.

U.S. Region	Year of Death				
	2002	2003	2004	2005	2006
Northeast	5,313	5,012	4,779	4,750	4,074
Midwest	1,726	1,666	1,482	1,378	1,325
South	7,337	7,427	7,535	7,907	6,475
West	2,571	2,585	2,600	2,234	2,141

(Data from: Centers for Disease Control and Prevention. *HIV/AIDS Surveillance Report, 2006.* Vol. 18. Atlanta: U.S. Department of Health and Human Services, Centers for Disease Control and Prevention; 2008: page 18. http://www.cdc.gov/hiv/topics/surveillance/resources/reports/.)

in columns and rows. All information in a single column is related, and all information in a single row is related. Look at the top of each column—these are column heads that tell you what information each column contains. Look at the beginning of each row. These are row heads, or labels that tell you what each row contains. In our example, the column heads reveal that the first column identifies each region, and the other columns contain the numbers of deaths for all regions for a given year. The row heads tell us that all deaths in a single row belong to a single region. So, by using the column and row heads, it is easy to find out, for example, how many deaths occurred in the Midwest in 2004.

TIME TO TRY

On a separate piece of paper, using sentences and paragraphs, write out all the information that is included in Table 2.4. Which version is easier to read? _____

GRAPHS

Graphs present a more pictorial view of data. Numerical data that can be organized in a table (**Figure 2.7a**) can usually also be presented in a graph. The main advantage is that the graph allows you to spot trends and relationships almost instantly. There are various types of graphs. Let's look at three of them: line graphs, bar graphs, and pie charts.

Graph title

Number of Measles Cases

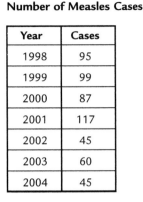

Year	Cases
1998	95
1999	99
2000	87
2001	117
2002	45
2003	60
2004	45

a)

Number of Measles Cases

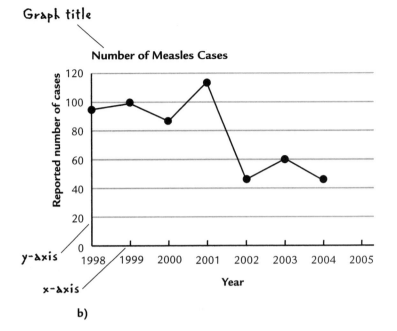

y-axis

x-axis

b)

FIGURE 2.7 Number of measles cases, 1998–2004. a) Table format. b) Basic line graph. Data From: CDC, *Summary of Notifiable Diseases 2003*, MMWR 52 (54) 4/22/05; MMWR 53 (52), 2005.

LINE GRAPHS   Look at **Figure 2.7b**. This is a line graph, and it is usually laid out in a grid. The horizontal axis at the bottom is the *x*-**axis**. It often, but not always, marks the progression of time. The vertical axis on the left is the *y*-**axis**, and it typically reflects some increasing value. Each axis should be labeled and should include units.

In our example, the *x*-axis tells us what years the data were collected. The *y*-axis gives the number of measles cases. For each year, the number of cases (data point) is placed above the year it represents. The data points may be left unconnected, they may be directly connected (as they are here), or a line of "best fit" may be drawn that passes between the points so they are evenly distributed on each side of it. In our example, because this depicts cases with time, the line allows us to see at once the cases for each year and to quickly comprehend the overall trend of a decrease in cases with time.

When drawing a line graph, don't forget to label the axes and to include units. The most common mistake made when graphing data is to not use an appropriate scale. Be sure to size the units to maximize the space the graph fills. You don't want the graph to be cramped into one corner, making it hard to read. Spread it out both vertically and horizontally. Also, be sure to give each graph or table an appropriate title.

BAR GRAPHS   **Figure 2.8** is a bar graph showing the changes in incidence of bacterial meningitis. Looking at the axes, you can see that the *x*-axis shows two separate years: 1985 and 1995. Each of these years has three bars—one for each of three bacterial species that cause meningitis. The *y*-axis tells what percentage of total cases each species represents. This graph is drawn so that the species data are directly compared by being positioned side by side, yet easily distinguished by use of different shading. The shading is explained below the graph in a feature called the **key**. Bar graphs may be drawn vertically or horizontally. Because each bar is so distinct, these graphs are good for comparing specific bits of data, whereas line graphs may be better at showing an overall trend.

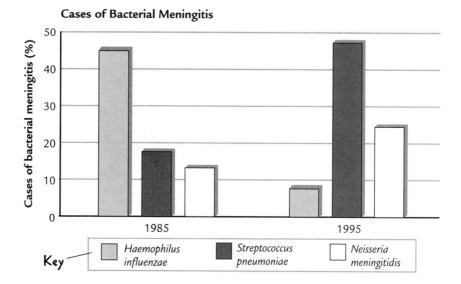

Cases of Bacterial Meningitis

Key: Haemophilus influenzae, Streptococcus pneumoniae, Neisseria meningitidis

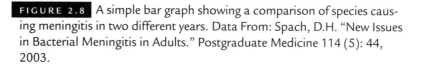

FIGURE 2.8   A simple bar graph showing a comparison of species causing meningitis in two different years. Data From: Spach, D.H. "New Issues in Bacterial Meningitis in Adults." Postgraduate Medicine 114 (5): 44, 2003.

PIE CHARTS   **Figure 2.9** is a pie chart, a type of graph that shows parts of a whole. This one shows the percentage of nosocomial (hospital-acquired) infections at various body sites. When looking at this, you immediately see that all the parts add up to the whole "pie," which is 100%, so these charts are effective when showing percentages. We all have a visual concept of a whole pie and a slice of pie, so even before looking at the numbers, we instantly see the differences in distribution. You know right away from this pie chart that one site (urinary tract) has the largest pecentage of the infections.

Pie charts have no axes to label, and the space within them is limited, so a key is often used. Labels for each "slice" may be written within the pie or placed to the outside, as in our example. Using different shading or colors for the slices makes these graphs even more readable. Because there are no axes to provide information, the graph title and key are very important.

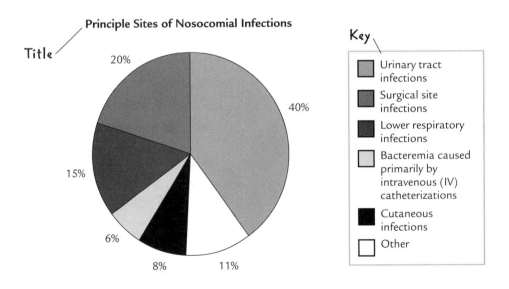

FIGURE 2.9   A pie chart showing sites of nosocomial infections. Data From: CDC, National Nosocomial Infection Surveillance.

When reading a graph, always read the title first and then the axes, key, and labels. Finally, just let your eyes take in the relationships depicted. ■

✔ QUICK CHECK

1. In Figure 2.7, during which year did the most cases occur?

_____

How many cases were there in 1999? _____

2. In Figure 2.8, which species was most common in 1985?

_____

3. In Figure 2.9, what is the second most common site for nosocomial infections? _____

Answers: 1. Most cases occurred in 2001, and there were 99 cases in 1999. 2. *Haemophilus influenzae* 3. surgical sites

Final Stretch!

Now that you have finished reading this chapter, it is time to stretch your brain a bit and check how much you learned.

WHAT DID YOU LEARN?

Try these exercises from memory first, and then go back and check your answers, looking up any items that you want to review. Answers to these questions are at the end of the book.

PART A: SOLVE THESE PROBLEMS.

1. $(4 \div 2) + 6 - 5 \times 2^3 =$ _____

2. $2 \times 10^4 =$ _____

3. 27/36 reduced is _____

4. $3/8 \times 2/3 =$ _____

5. $5/6 - 7/12 =$ _____

6. $0.5 \times 0.4 =$ _____

7. If a water sample has 1,000 bacteria per milliliter, how many does it have per liter? _____

8. 4 meters = _____ centimeters

9. 1 inch = 2.54 cm, so 1 foot = _____ cm

10. If 18% of a hospital's 350 patients had an infection caused by a methicillin-resistant strain of *Staphylococcus*, how many patients have this infection? _____

PART B: ANSWER THESE QUESTIONS.

1. What is the mean of 8, 9, 12, 18, and 23? _____

2. What is the numerator in 4/5? _____

3. Express 3/10 as a decimal _____ and as a percent _____

4. From Figure 2.9, what percentage of nosocomial infections are in the lower respiratory tract? _____

5. What would you be measuring if you are looking at a meniscus? _____

6. In the metric system, list the base unit for each of the following:
   mass: _____
   length: _____
   volume: _____

7. If the ratio of water to sugar in a solution is 4:1, what does this mean? _____

8. At what Celsius temperature does water boil? _____

9. The amount of space something occupies is called _____.

10. How many milligrams are there in 1 gram? _____

**PART C: WRITE THE CHAPTER TERMS IN A NOTEBOOK AND
DEFINE THEM IN YOUR OWN WORDS. GO BACK THROUGH
THE CHAPTER TO CHECK YOUR MEANINGS, CORRECTING
THEM AS NEEDED. LIST EXAMPLES WHEN APPROPRIATE.**

Data
Product
Exponents
Dividend
Divisor
Quotient
Normal
Average
Mean
Numerator
Denominator
Equivalent fractions

Reduce
Greatest common factor
  (GCF)
Common
  denominator
Least common
  multiple (LCM)
Repeating decimal
Scientific notation
Coefficient
Ratio
Metric system

International System of Units
Meter (m)
Gram (g)
Liter (L)
Degree Celsius (°C)
Volume
Meniscus
x-axis
y-axis
Key

3 Terminology

The ABCs of Microbiology

When you complete this chapter, you should be able to

- Break down biological terms to understand their meanings.

- Build biological terms from basic word roots, prefixes, and suffixes.

Your Starting Point

Answer the following questions to assess your knowledge of biological terminology.

1. Most biological terms arise from which languages? _____

_____

2. What is the difference between an *abbreviation* and an *acronym*? _____

_____

3. What is meant by the term *hydrophobic*? _____

_____

4. What is the difference between a *prefix* and a *suffix*? _____

_____

5. What is an *eponym*? _____

_____

How Do You Say, In Your **Language** . . .?

Have you ever read a computer manual only to find that you are still unsure of how to configure your personal firewall, or for that matter why you should? Has your mind gone numb as an auto mechanic explained all those expensive malfunctioning parts that were replaced to get rid of that mystery noise in your car? Have you tried to read the small print in an advertisement for a new prescription drug? At such

Answers: 1. Latin and Greek. 2. An abbreviation is a shortened form of a word or phrase; an acronym is a word formed from the first or key letters of each word in a full name. 3. *Hydro-* means "water," and *-phobic* means "fearing"; this term refers to substances such as oils that do not mix with water. 4. A prefix is a word part added at the beginning of a word, whereas a suffix is added at the end. 5. Eponyms are terms that include someone's name.

times, it may seem as if other people speak a secret language. Indeed, most professions have their own sublanguage, as do most academic disciplines. Microbiology is no exception.

Learning the terminology in this discipline may be a challenge, but we won't say it is difficult. *Difficult* leaves the option of ducking behind excuses. You should think of the words as a challenge because a challenge sparks the competitor in us—it makes us try harder. Although the words may seem intimidating at first, you will quickly learn tools and tricks to help you understand even the longest terms.

Obviously if you plan to spend time in a place where another language is spoken, communication could be a challenge unless you learn some of the local language in advance. Your microbiology class may seem like a place where another language is spoken. Much of the terminology will be new, and most of the words do, in fact, come from languages other than English. You will learn about microbes by listening to lectures filled with this new language, and you will be expected to discuss your course material and write exam answers using this new language. That is why it is so important that you learn the language of microbiology. As you study, the first thing you should do is master—not just read, but *master*—the new words. You have to know the language before you can understand and join in the conversation.

Learning the vocabulary is the first step in learning microbiology. ▪

We recommend that students maintain a running vocabulary list. The easiest way to do this is to keep a separate notebook into which you write all new terms as they are introduced (**Figure 3.1**). If you write new terms in your notes during lecture, transfer them to your vocabulary list and be sure to check their spellings later. Once you know that a term is spelled correctly, write the actual definition—exactly what does the word mean? You can find the actual textbook definition in your textbook or in a biological dictionary, but you should also try to explain the term in your own words.

Next, be sure you can use the word properly in a sentence. Try to add some examples that illustrate the term, if appropriate. For example, a microorganism is any organism that is not visible with the unaided eye. Examples of microorganisms include bacteria, protozoans, and yeast. If

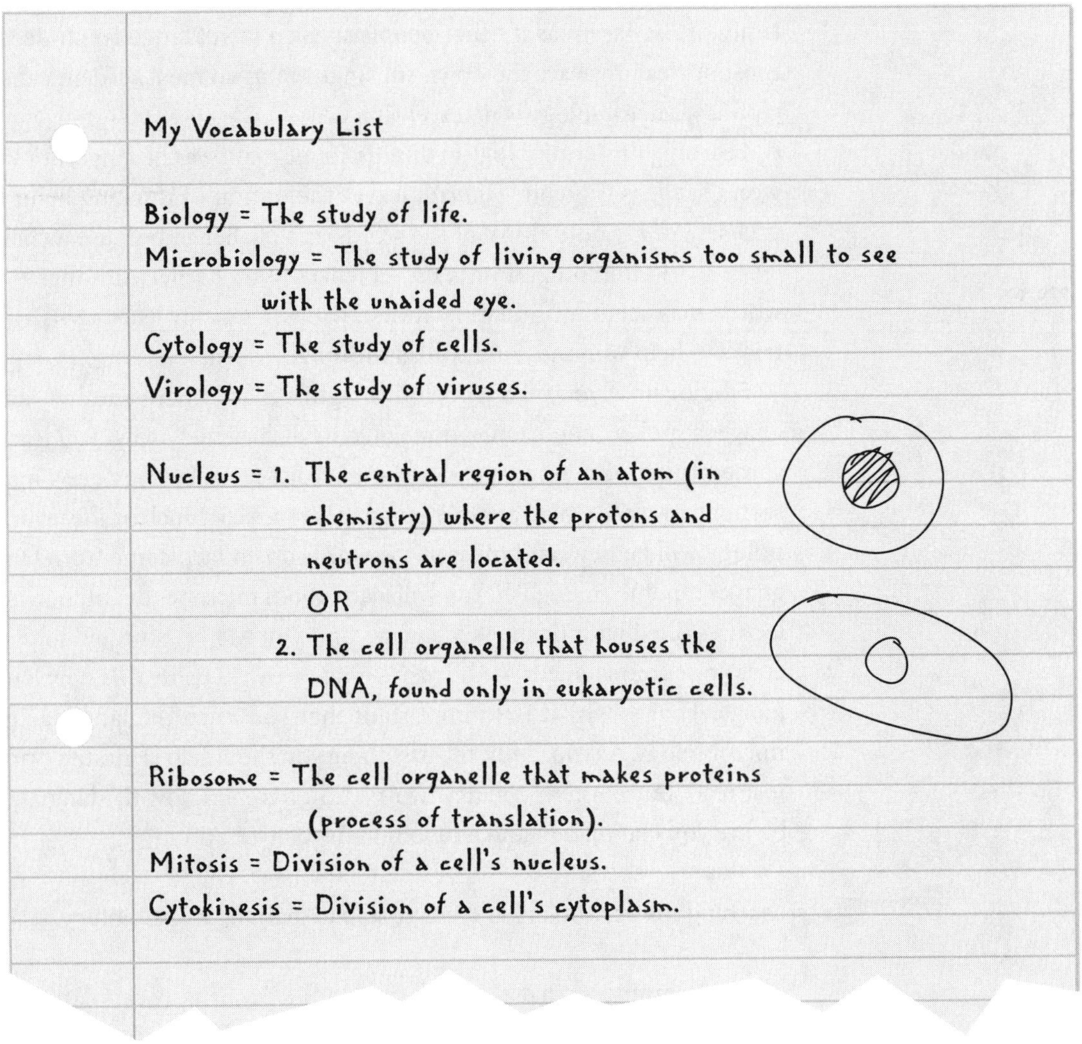

My Vocabulary List

Biology = The study of life.
Microbiology = The study of living organisms too small to see
                        with the unaided eye.
Cytology = The study of cells.
Virology = The study of viruses.

Nucleus = 1. The central region of an atom (in
                    chemistry) where the protons and
                    neutrons are located.
                OR
              2. The cell organelle that houses the
                    DNA, found only in eukaryotic cells.

Ribosome = The cell organelle that makes proteins
                    (process of translation).
Mitosis = Division of a cell's nucleus.
Cytokinesis = Division of a cell's cytoplasm.

FIGURE 3.1 Keeping a separate running vocabulary list throughout the semester can help you master the language.

you are a visual learner, try illustrating your new words if you can. If you are an auditory learner, try reading the words and their meanings out loud and consider recording your list. If you are a tactile learner, you might benefit from writing the terms and their meanings on flash cards.

You may be surprised at the beginning of the course at the number of new terms you encounter. You can think of the words as a kind of smoke screen—the underlying principles of microbiology are rather simple, but you may not see that through the smoke. More than one student has jokingly commented that "It's all Greek to me." They don't realize that their joke is not far from the truth—most biological terms have either Greek or Latin origins, although terms may be derived from other languages as well. So, instead of thinking that it is hard to learn microbiology because of all the big words, look at it this way: While learning microbiology, you will learn some new languages!

✔ **QUICK CHECK**

From which languages do most biological terms originate?

Answer: Greek and Latin.

The More the Merrier . . . **Types of Terms**

When studying microbiology, you will encounter many different types of terms, many of which you will already know.

DESCRIPTIVE TERMS

Most terms you will encounter are descriptive, meaning the name is closely related to its meaning. For most students, these words often look the most intimidating—they can be quite long. Yet, once you know some simple rules, **descriptive terms** become quite easy to understand and deciphering their meanings actually becomes fun. Even large words like *chemolithoautotroph* become manageable with some information and practice (see **Table 3.1** on page 91), and just think how empowered you will feel when you can actually converse with scientists or physicians in their own language! We will spend most of our time in this chapter working on descriptive terms.

EPONYMS

Another type of term is the **eponym**, which literally means "putting a name upon." Eponyms are terms that include someone's name and have been used traditionally to honor the person who first described a certain structure or condition. This practice led to the following terms:

■ Golgi apparatus, a structure with the specialized function of packaging and transporting substances within some eukaryotic cells. These were named for Camillo Golgi, an Italian cell biologist.

■ Pasteurization, a method currently used today for controlling growth of bacteria in dairy products, developed by Louis Pasteur in nineteenth-century France to prevent spoilage of wine.

■ Gram stain, a method to stain and distinguish bacteria based on their cell wall thicknesses, developed by Danish scientist Hans Christian Gram in the late nineteenth century.

Obviously, these terms are not so easily understood. Few people now associate an eponym with the discoverer, making the terms harder to master. Recent practice has moved away from eponyms to a preference for descriptive terms, so those will be our focus. You will learn relevant eponyms as you move through your course, but fewer and fewer are in use today. You will occasionally encounter them, and you just need to memorize them as they come up.

TIME TO TRY

Hansen's disease is also called *leprosy.*

1. Which is the descriptive term? _____

2. Which is the eponym? _____

3. With which term are you more familiar? _____

4. The word *leprosy* originated from the Greek word *lepros* which means "scaly." Can you guess what one of the characteristics of leprosy might be? _____

_____

Therefore, *leprosy* is the descriptive term. In this case, the term *leprosy* often has a negative connotation, dating way back to biblical times when people with the disease (lepers) were considered contagious and forced to live a solitary life away from their communities. So, the eponym, Hansen's disease, is often used in medicine to avoid the negative connotation.

ABBREVIATIONS AND ACRONYMS

Like many disciplines, the world of microbiology is full of shortcuts. After all, some of those descriptive terms grow quite large! You will often encounter modifications of full terms in the form of abbreviations or acronyms. You are already familiar with abbreviations. An **abbreviation** is a shortened form of a word or phrase. The word *abbreviation* itself can be abbreviated *abbrev.*

An **acronym** is technically a word formed from the first letter or key letters of each word in a multiple–word term. It is pronounced as if it were a word. For example, AIDS is the acronym for <u>A</u>cquired <u>I</u>mmune <u>D</u>eficiency <u>S</u>yndrome and is pronounced like the word *aids.*

Sometimes abbreviations are formed like an acronym by using only certain letters from multiple words. For example, HIV stands for <u>h</u>uman <u>i</u>mmunodeficiency <u>v</u>irus, the cause of AIDS. But we pronounce the acronym, AIDS, as a word, while pronouncing each letter in the abbreviation—H-I-V[*].

When using abbreviations and acronyms, always be sure that you know the full names and their meanings and that you list the letters in the correct order. ∎

---

[*] Note that some sources consider HIV an acronym; others consider it an initialism (another type of abbreviation), while others simply lump abbreviations and acronyms together.

TIME TO TRY

Let's see how many of these abbreviations and acronyms you either know or can figure out. Match the following abbreviations to their full names.

1. _____ STD   a) Severe acute respiratory syndrome

2. _____ RNA   b) Adenosine triphosphate

3. _____ SARS   c) Ribonucleic acid

4. _____ ATP   d) Sexually transmitted disease

It should be easy for you to find the answer with the correct letters in it.

✔ **QUICK CHECK**

1. What makes an acronym different from an abbreviation?

_____

2. HAART stands for highly active antiretroviral therapy, a type of treatment for people with HIV. Is HAART (pronounced "heart") an acronym or abbreviation? _____

Answers: 1. An acronym is an actual word formed from the letters in an abbreviation. 2. acronym

Putting Down Roots and **Building Descriptive Terms**

Most biological terms are descriptive and are built from two or more of three basic parts:

■ a **prefix**, at the beginning of the word,

■ a **root**, which is the main focus of the word, and

■ a **suffix**, at the end of the word.

New words may be made simply by changing these parts (**Figure 3.2**). Some time ago, when learning English, you learned about prefixes

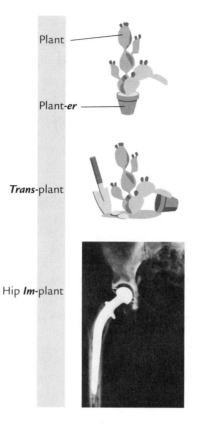

Plant

Plant-*er*

Trans-plant

Hip *Im*-plant

FIGURE 3.2 Adding prefixes and suffixes to the root "plant" will change the meaning.

and suffixes and how to use them to change one word to another, and you now use them all the time. You will be amazed how many word roots, prefixes, and suffixes you already know.

A prefix is a short addition placed at the front of a word root. Consider the root *cycle*, which means circle or wheel. We can change the specific meaning of this root by adding prefixes as follows:

- uni*cycle*—has one wheel

- bi*cycle*—has two wheels

- tri*cycle*—has three wheels

- motor*cycle*—has a motor

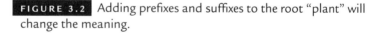

You likely knew the prefixes we just used in our example. When you are reading biological terms, don't let the first glance worry you. Simply take a breath and look carefully at the word. Find the root and then examine any prefixes or suffixes that are used with it. Sometimes you may not know a particular word part at first, but you may figure it out by thinking of words you *do* know that include that part. Let's try some.

TIME TO TRY

Here are some word parts and examples of words that use them. For each of the following, what do you think the word part means? Look at the examples and see if there is something they all share in common. This technique can help you determine the meaning of words with which you are unfamiliar—first look for the word parts and think of other terms that you do know that contain them.

Word part	Examples	Meaning
Root: bio-	Biology, microbiology, biosphere	_____ _____
Prefix: micro-	Microbiology, microscope, micrometer	_____ _____
Suffix: -ology	Microbiology, parasitology, mycology	_____ _____

Let's look at the name of this course: microbiology. The root *bio-* means "life," the prefix *micro-* means "very small," and the suffix *–ology* means "the study of." Microbiology is therefore the study of very small living things. Now that you know the meaning of the root *bio-*, you can decipher the meaning of biology (the study of life) and biosphere (the living organisms on Earth). Once you know the meaning of the prefix *micro-*, you can figure out that a microscope allows you to look at tiny things, and a micrometer is a very small measurement. The other two areas of study mentioned above are *parasitology*, which is the study of parasites, and *mycology*, which is the study of fungi.

Not only do descriptive terms convey a clear meaning, but they are easy to translate once you know the prefixes, suffixes, and roots. You already know several of these, and you will learn plenty more as you go. To help jump-start your language acquisition, **Table 3.1** lists many word parts commonly used in microbiology, along with their meanings and an example of each. Don't worry about learning them all now—they are there to help you start building your language skills. If you get stuck on a word that appears in your textbook or that your instructor uses in a lecture, use this table as a reference. You should read through the terms and see which ones you already recognize.

Mastering word roots, prefixes, and suffixes is the most effective way to expand your microbiology vocabulary because they can be recombined into countless new words. ■

TABLE 3.1   Some common word parts used in microbiology.

Word part	Meaning	Example
a-	not, without	acellular: has no cells
abyss-	deep, bottomless	abyssal zone: the cold, dark benthic communities near the bottom of the ocean
ad-	toward	adhesin: molecule that helps a bacterium attach to its host
aero-	air	aerobic: referring to chemical reactions using oxygen
agglutinat-	glued together	agglutination: an immune response in which bacteria or viruses are clumped together
allo-	different	allograft: transplantation of tissue from a donor to a genetically dissimilar recipient
alveol-	a cavity	alveoli: one of the dead-end, multilobed air sacs that constitute the gas exchange surface of the lungs
amphi-	dual	amphipathic molecule: a molecule that has both a hydrophobic and a hydrophilic region
an-	not, without	anaerobe: an organism that cannot tolerate oxygen
ana-	up	anabolic pathway: a metabolic pathway that consumes energy to build up complex molecules from simpler ones
anti	against	antibody: a protein that attaches to a specific kind of antigen and helps counter its effects; part of an immune response
-aphy	suck	anaphylactic shock: an acute, life-threatening allergic response where breathing becomes difficult

▶

TABLE 3.1   Some common word parts used in microbiology, continued.

Word part	Meaning	Example
apo-	off, away	apoptosis: cell suicide
aqua	water	aqueous humor: the clear, watery solution that fills the anterior cavity of the eye
arch-	ancient, beginning	Archaea: a domain of bacteria that was once thought to be the most primitive or ancient
arthro-	jointed	Arthropoda: segmented invertebrates with exoskeletons and jointed appendages
asco-	sac	Ascomycota: a division of fungi with saclike fruiting bodies
auto-	self	autotroph: an organism that "feeds" itself by making carbohydrates from carbon dioxide
aux-	grow, enlarge	auxotroph: a member of a species that has atypical nutritional requirements for growth
bacil	a rod	bacillus: a rod-shaped bacterium
baro-	pressure	barophile: an organism that can tolerate high hydrostatic pressure environments
basidio-	club	Basidiomycota: a division of fungi that have club-shaped fruiting bodies
baso-	basic	basophil: a cell that has an affinity for basic stains
bentho-	the depths of the sea	benthic zone: the bottom surfaces of aquatic environments
bi-	two	binomial: two names
bin-	two at a time	recombinant: an offspring whose phenotype differs from that of the parents binary fission: reproduction of a prokaryotic cell into two cells
bio-	life	biology: the scientific study of life biosphere: all the environments on Earth that are inhabited by life antibiotic: a chemical that inhibits microbial growth
-bios	life	symbiosis: an ecological relationship between organisms of two different species that live together in direct contact
capsa	a box	capsid: the protein shell that encloses the viral genome capsule: the polysaccharide coating on the surface of some bacterial cells
carb-	coal	carboxyl group: a functional group present in organic acids, consisting of a carbon atom double-bonded to an oxygen atom and a hydroxyl group
cardi-	heart	cardiovascular system: the closed circulatory system characteristic of vertebrates
cata-	down	catabolic pathway: a metabolic pathway that releases energy by breaking down complex molecules into simpler ones

TABLE 3.1 Some common word parts used in microbiology, continued.		
Word part	**Meaning**	**Example**
centro-	center	centromere: the narrow "waist" of a condensed replicated chromosome, where two sister chromatids are attached
cephal-, cephalo-	head	encephalitis: an inflammation of the brain
chemi-	chemical	chemiosmosis: the production of ATP using the energy of hydrogen ion gradients across membranes to phosphorylate adenosine diphosphate (ADP)
chemo	chemical	chemotroph: an organism that is able to use chemicals for energy chemotaxis: movement toward or away from a particular chemical
chloro-	green	chloroplast: the site of photosynthesis in plants and algae
chol	anger	cholera: a disease characterized by vomiting and diarrhea
chroma-	colored	chromatin: DNA and the various associated proteins that form eukaryotic chromosomes
chryso-	golden	Chrysophyta: a division of algae that have golden or yellow-green pigments
cili-	hair	cilium: a short hairlike cellular extension with a microtubule core
-clin	slope	thermocline: a narrow stratum of rapid temperature change in the ocean and in many temperate-zone lakes
co-	together	cofactor: an ion or molecule that is essential for an enzyme to function properly coinfection: a condition in which a patient is infected simultaneously with two different organisms
coeno-	common	coenocytic: a multinucleated condition resulting from the repeated division of nuclei without cytoplasmic division
com-	together	recombinant: an offspring whose phenotype differs from that of the parents
con-	with, together	condensation reaction: a chemical reaction in which two molecules become covalently bonded to each other through the loss of a small molecule, usually water
conjug	together	conjugation: in bacteria, the transfer of DNA between two cells that are temporarily joined
cortico-	the shell	corticosteroids: hormones produced by outer layers of glands; they suppress the immune system
counter-	opposite	counterstain: a second, contrasting stain
cuti, -cutane	the skin	subcutaneous injection: injection beneath the skin
cyano-	blue	cyanobacteria: a group of photosynthetic bacteria that are blue-green
cycl-	circle	cyclic photophosphorylation: a method of photosynthetically producing ATP that ends where it begins

▶

TABLE 3.1   Some common word parts used in microbiology, continued.

Word part	Meaning	Example
cyst	sac, bladder	blastocyst: a hollow ball of cells produced 1 week after fertilization in humans
-cyte	cell	leukocyte: white blood cell
cyto-	cell	cytosol: the semifluid medium in a cell in which all internal components are located
de-	from, down, out	denitrification: the process of converting nitrate back to nitrogen
dendro-	tree	dendritic cell: a cell with numerous branches, which devours pathogens
-derm	skin	epidermis: the outermost layer of the skin
di-	two	diploid: cells that contain two homologous sets of chromosomes
-dilat	expanded	vasodilation: an increase in the diameter of blood vessels, allowing more blood to flow through them
dino	terrible	dinoflagellate: a unicellular flagellated protist, often with armorlike plates
diplo-	double	diplococcus: a pair of spheres
dis-	separate	nondisjunction: an accident of meiosis or mitosis, in which both members of a pair of homologous chromosomes or both sister chromatids fail to separate
dys-	bad, difficult	dyspnea: difficulty in breathing dysentery: disease characterized by severe diarrhea
eco-	house	ecosystem: a biological community and the physical environment associated with it
electro-	electricity	electroporation: a technique used to introduce recombinant DNA into cells by applying a brief electrical pulse to a solution containing those cells
-ell	small	organelle: a small formed body with a specialized function found in cells
-emia	in the blood	septicemia: condition in which pathogens in the blood are causing illness
encephal-	in the brain	encephalitis: inflammation in the brain
end-, endo-	within, inner	endemic: a disease that occurs at a stable frequency within a population endotoxin: a toxin contained within the bacterial cell wall and released when the cell dies
entero-	intestine, gut	Enterovirus: a virus that infects the intestines
epi-	above, over	epidemic: a disease that occurs with a greater-than-normal frequency
erythro-	red	erythrocyte: red blood cell
estuar-	the sea	estuary: the area where a freshwater stream or river merges with the ocean

TABLE 3.1   Some common word parts used in microbiology, continued.

Word part	Meaning	Example
etio-	cause	etiology: study of the cause of disease
eu-	true, good	eukaryotic cell: a cell that has a true nucleus
ex-, exo-	out, outside	exocytosis: the movement of materials out of a cell
		exotoxin: a toxin secreted by a bacterium into its environment
extra	outside	extracellular digestion: the breakdown of food outside cells
fertil	fruitful	fertilization: the union of gametes to produce a zygote
flagell-	whip	flagellum: a long whiplike cellular extension that moves cells
-form	shape	transformation: the process that converts a bacterium into a new form
gam-	marriage	monogamous: a type of relationship in which one male mates with just one female
gamet-	wife or husband	gamete: an egg or sperm cell
gamy	reproduction, marriage	syngamy: the process of cellular union during fertilization
gastro-	stomach, belly	gastroenteritis: inflammation of the stomach and intestines
gen-, gene-	produce	genome: a cell's endowment of deoxyribonucleic acid (DNA)
		polygenic: an additive effect of two or more gene loci on a single phenotypic character
genesis	origin, birth	biogenesis: living organisms are produced from other living organisms
-genic	producing	pathogenic: disease causing
geno-	offspring	genotype: the genetic makeup of an organism
-geny	origin	phylogeny: the evolutionary history of a related group of organisms called a taxon
gest	carried	ingestion: a heterotrophic mode of nutrition in which other organisms or detritus are eaten
-glob	globe, sphere	immunoglobulin: one of the class of proteins that act in an immune response
glyco-	sweet	glycocalyx: the sticky polysaccharide covering of some bacteria
-graphy	writing	demography: the study of statistics relating to births and deaths in populations
halo-	salt	halophile: microorganisms that live in highly saline environments, such as the Great Salt Lake or the Dead Sea
haplo-	single	haploid: cells that contain only one chromosome of each homologous pair
helic-	a spiral	helicase: an enzyme that untwists the double helix of DNA at the replication forks
hemo-	blood	hemolysis: breakdown of hemoglobin molecules in red blood cells

▶

TABLE 3.1	Some common word parts used in microbiology, continued.

Word part	Meaning	Example
hepa-	liver	hepatitis: inflammation of the liver
hetero-	other, different	heterozygous: an organism or cell that has two different alleles at a given gene location on homologous chromosomes
hist-	tissue	histamine: an inflammatory chemical released from damaged tissues
holo-	whole	holoenzyme: the apoenzyme and cofactor
homeo-	same	homeostasis: maintenance of a stable internal environment of the body
homo-	alike	homozygous: an organism or cell that has two identical alleles at a given gene locus on homologous chromosomes
horm-	to set in motion, excite	hormone: in multicellular organisms, one of many types of circulating chemical signals
hydro-	water	hydrocarbon: an organic molecule consisting only of carbon and hydrogen
hyper-	over, above, excessive	hypersensitivity: an overreaction of the immune system, as with allergies
hypo-	under, below, lower	hypotonic: having a lower solute concentration
immun	safe, free	immunization: the process of generating a state of immunity to disease by artificial means
inter-	between	interleukin: a molecule used for communication between leukocytes
intra-	inside, within	intracellular digestion: the joining of food vacuoles and lysosomes to allow chemical digestion to occur within the cytoplasm of a cell
intro-	within	intron: a noncoding, intervening sequence within a eukaryotic gene
iso-	equal	isotope: one of two or more atoms of the same element having the same number of protons and electrons but a different number of neutrons
-itis	inflammation	hepatitis: inflammation or infection of the liver
karyo-	nucleus	karyotype: a display of the chromosomes of a cell arranged in pairs
kilo-	a thousand	kilocalorie: a thousand calories
kine, kinet, -kine	movement	cytokine: a molecule used to send messages between cells kinetic energy: the energy of motion
lact-	milk	lactose: sugar found in milk
lepro-	scaly	leprosy: a chronic disease that affects the skin and peripheral nerves
leuko-	white	leukocyte: a white blood cell
liga-	bound or tied	ligand: a small molecule that specifically binds to a larger one
limn-	a lake	limnetic zone: the well-lit, open surface waters of a lake farther from shore
lipo-	fat	lipopolysaccharide: a molecule that makes up a Gram-negative outer membrane, made of lipid and polysaccharide
litho-, -lith	stone, rock	lithotroph: an organism that obtains energy from inorganic molecules

| TABLE 3.1 | Some common word parts used in microbiology, continued. | |

Word part	Meaning	Example
littor-	seashore	littoral zone: the shallow, well-lit waters of a lake close to shore
lopho	crest, tuft	lophotrichous: an organism with flagella in a tuft
lut-	yellow	*Micrococcus luteus*: a bacterium with a yellow pigment
-lyse	break	hydrolyse: to break chemical bonds by adding water
-lysis	split	glycolysis: the splitting of glucose into pyruvate
lyso-	loosen	lysosome: a membrane-bounded sac of hydrolytic enzymes that a cell uses to digest macromolecules
lyt-	loosen	lytic cycle: a type of viral replication cycle resulting in the release of new viruses by death or lysis of the host cell
macro-	large	macromolecule: a large molecule
magnet-	magnetic	magnetosome: iron-containing structure orients bacteria along a geo-magnetic field
mal-	bad or evil	malignant tumor: a cancerous tumor that is invasive enough to impair the functions of one or more organs
meio-	less	meiosis: a variation of cell division that yields haploid daughter cells, which have half as many chromosomes as their diploid parent cell
meso-	middle	mesophile: organisms that require moderate temperatures (20°C–40°C)
meta-	boundary, turning point; change; with; between	metaphase: the mitotic stage in which the chromosomes are aligned in the middle of the cell, at the metaphase plate
metr	measure	metric: a form of measurement unit
micro-	small	microbiology: the study of small organisms
mito-	a thread	mitosis: the process of nuclear division in eukaryotic cells, in which the threadlike replicated chromosomes are allocated equally to the two daughter nuclei produced
mono-	one	monotrichous: an organism having one flgellum
morph-, morpho-	form	morphology: shape
muta-	change	mutation: a change in the DNA of a gene
mutu-	reciprocal	mutualism: a symbiotic relationship in which both the host and the symbiont benefit
myco-	fungus	mycorrhizae: mutualistic associations of plant roots and fungi
necro-	death	necrosis: cell death
neutr, neutro-	neutral	neutrophil: a white blood cell that stains equally with acid or basic dyes
nom-	name	binomial: a two-part latinized name of a species
non-	not	nonpolar bond: a molecule in which electrons are equally shared between atoms; neither atom takes the electrons

▶

TABLE 3.1		Some common word parts used in microbiology, continued.

Word part	Meaning	Example
nucle-, nucleo-	nucleus	nucleoid: the region where the genetic material is concentrated in prokaryotic cells nucleosome: the basic beadlike unit of DNA packaging in eukaryotes
-oid	like, form	nucleoid: the region where the genetic material is concentrated in prokaryotic cells
ology	the study of	microbiology: the study of microorganisms
onco-	tumor	protooncogene: a normal cellular gene with the potential to cause cancer if it is altered to become an oncogene
-osis	a condition of	mycosis: the general term for a fungal infection
osmo-	pushing	osmosis: the diffusion of water across a selectively permeable membrane
pan-	all, everything	pandemic: an epidemic that occurs worldwide
para-	beside, near	parasite: an organism that lives in or on another organism
parvo-	small	parvoviruses: a group of extremely small viruses
patho-	disease	pathogen: a microorgansim that causes disease
pelag-	the sea	oceanic pelagic biome: most of the ocean's waters far from shore, constantly mixed by ocean currents
perfora-	bore through	perforin: a protein that forms pores in a target cell's membrane
peri	around	periodontal disease: a disease that involves infection of the tissues surrounding the teeth
phaeo-	brown	Phaeophyta: division of brown-pigmented algae
-phage	to eat	bacteriophages: viruses that infect bacteria
phago-	to eat	phagocytosis: cell eating; the process by which food or foreign particles are enveloped by cells called phagocytes
pheno-	appear	phenotype: the physical and physiological traits of an organism
-phil	loving	neutrophil: the most abundant type of leukocyte, neutrophils destroy foreign invaders and release various substances
-philos	loving	hydrophilic: having an affinity for water halophile: an organism that likes salty environments
-phobos	fearing	hydrophobic: having an aversion to water
-phore	bearer	siderophore: a molecule used by bacteria to bring iron into their cells
photo	light	phototroph: an organism that uses light as a source of energy
phyco-	algae	phycology: the study of algae
-phyll	leaf	chlorophyll: photosynthetic pigment in cyanobacteria and green algae

TABLE 3.1   Some common word parts used in microbiology, continued.

Word part	Meaning	Example
phylo-	tribe	phylogeny: the evolutionary history of a group of related organisms called a taxon
phyto-, -phyta	plant	phytoplankton: tiny photosynthetic organisms that live near the surface of the ocean Chlorophyta: division of algae that have green pigments
pino-	drink	pinocytosis: cell drinking; the process by which a living cell engulfs a droplet of liquid
plasm-, plasmo	molded	plasmolysis: a phenomenon in walled cells in which the cytoplasm shrivels, and the cytoplasmic membrane pulls away from the cell wall when the cell loses water to a hypertonic environment
pleio-	more	pleiomorphic: a bacterial species that can have many shapes
pneumo-	air, lung	pneumonia: inflammation of the lungs caused by bacteria or viruses
-pod, -podium	foot	pseudopodium: a cellular extension of amoeboid cells used in moving and feeding
poly	many	polymer: a large molecule made of multiple copies of a monomer
-port	gate, door	cotransport: the coupling of the "downhill" diffusion of one substance to the "uphill" transport of another across a membrane and against its own concentration gradient
post-	after	postpolio syndrome: a condition that affects people many years after having had polio as a youth
pre-	before, first	precursor metabolite: a molecule that is used for synthesizing other molecules
pro-	before, first	prodromal period: the early stage of a disease in which there are only mild symptoms
profund-	deep	profundal zone: the deepest region of a lake, where no light penetrates
pseudo	false	pseudogenes: DNA segments very similar to real genes but which do not yield functional products
pulmo-	a lung	pulmonary histoplasmosis: a fungal infection of the lungs with symptoms similar to tuberculosis
quadr-	four	quadruped: any animal having four feet
re-	again	recombinant: an offspring whose phenotype differs from that of the parents
regula	regular	osmoregulation: adaptations to control the water balance in organisms living in hyperosmotic, hypoosmotic, or terrestrial environments
retro	backward	retrovirus: an RNA virus that reproduces by transcribing its RNA into DNA and then inserting the DNA into a cellular chromosome

►

TABLE 3.1   Some common word parts used in microbiology, continued.

Word part	Meaning	Example
rhizo-	root	rhizoids: long tubular single cells or filaments of cells that anchor bryophytes to the ground
rhodo	red	Rhodophyta: a division of algae that are characterized by red pigments
-sacchar	sugar	monosaccharide: simplest type of sugar
schizo	split	schizont: a form of a parasite that undergoes multiple mitoses and then cytokinesis to form many cells
script	write	transcription: the synthesis of RNA on a DNA template
semi-	half	semisynthetic antimicrobial: an antimicrobial drug that has a natural origin but has been chemically altered
sep-, sept-	putrefaction	aseptic: a technique used to prevent contamination by pathogens
sphero-, -sphere	ball-like	biosphere: the portion of Earth that is inhabited by living organisms
spir-, spiro-	helical	spirochete: a helical-shaped bacterium
sporo-	seed	sporophyte: the multicellular diploid form in organisms such as plants that undergo alternation of generations that results from a union of gametes and that meiotically produces haploid spores that grow into the gametophyte generation
staphylo-	cluster, bunch	staphylococcus: a cluster of spherical-shaped bacteria
-stasis	standing, staying	stationary phase: a part of a bacterial growth curve in which the number of cells remains constant
strepto-	chain	streptococcus: spherical bacteria arranged in a chain
stromato	something spread out	stromatolite: a rocklike structure composed of layers of prokaryotes and sediment
sub-	under, beneath	subclinical: absence of symptoms even though clinical tests indicate signs of the disease
sulf-	sulfur	sulfhydryl group: a functional group that consists of a sulfur atom bonded to an atom of hydrogen
supra-, super-	above, over	superinfection: a patient with one infection becomes infected with a second pathogen
sym-	with, together	symbiosis: a close association between two organisms
syn-	together	syncytium: a giant cell with many nuclei, often formed from fusion of virus-infected cells synthesis: combining separate elements into a single, unified whole
-taxis	movement	chemotaxis: movement toward or away from a chemical
taxo-	arrange	taxonomy: the branch of biology concerned with naming and classifying the diverse forms of life

TABLE 3.1   Some common word parts used in microbiology, continued.

Word part	Meaning	Example
telos-	an end	telomere: the protective structure at each end of a eukaryotic chromosome
tetan	rigid, tense	tetanus: *Clostridium tetani* produces a toxin that causes muscles to fail to relax.
tetra-	four	tetrad: a group of four spherical bacteria that remain attached in a single plane following cell division
thallos-	sprout	thallus: a body of an alga or a fungus that is plantlike but lacks true roots, stems, and leaves
therm-, thermo	heat, temperature	thermophile: a microorganism that requires high temperatures (above 45°C)
thio-	sulfur	thiol: organic compounds containing sulfhydryl groups
thylaco-	sac or pouch	thylakoid: a series of flattened saclike membranous structures within chloroplasts
-tonus	tension	hypertonic: referring to a solution with a higher concentration of solutes in comparison to another solution
-topo	place	epitope: a small region on the surface of an antigen to which an antigen receptor or antibody binds
trans-	across	transduction: transfer of DNA from one bacterium to another by a bacteriophage
tri-	three	trimester: a 3-month period
-troph	food, nourishment	photoautotroph: an organism that harnesses light energy to drive the synthesis of organic compounds from carbon dioxide
tropho-	food, nourishment	trophozoite: the motile feeding stage of a protozoan
-tubul	a little pipe	microtubules: hollow rods of protein found in the cytoplasm of almost all eukaryotic cells
ultra-	beyond	ultracentrifuge: a machine that spins test tubes at the fastest speeds to separate liquids and particles of different densities
uni-	one	unicellular: single-celled organisms
vacu-	empty	vacuole: sac that buds from the endoplasmic reticulum, Golgi apparatus, or plasma membrane in eukaryotic cells
vagin-	a sheath	invagination: the infolding of cells
-valent	strength	covalent bond: an attraction between atoms that share one or more pairs of outer-shell electrons
vasa, vaso	vessel	vasoconstriction: narrowing of the diameter of a blood vessel

▶

TABLE 3.1   Some common word parts used in microbiology, continued.

Word part	Meaning	Example
vect-, -vect	carried	vector: an arthropod that transmits disease from one host to another
villi	shaggy hair	microvilli: many fine, fingerlike projections of the epithelial cells in the lumen of the small intestine that increase its surface area
virul	poisonous	virulent virus: a virus that reproduces only by a lytic cycle, thus causing the death of the host cell
xantho-	yellow	xanthophyll: one of a class of yellow pigment molecules present in some algae
xeno-	strange, foreign	xenograft: transplant of tissue from one species into a different species axenic: having only one organism present
-yl	substance or matter	adenylyl cyclase: an enzyme built into the plasma membrane of cells that converts ATP to cyclic AMP
-zoa, zoan, zoo-	animal	zoonosis: a disease that is spread from animal to human
zygo-	yoke, joining	zygote: the cell resulting from the union of two gametes

TIME TO TRY

Now you can practice putting your new knowledge to use. Match the following terms with their meanings.

1. _____ arthritis

2. _____ bacteremia

3. _____ phototroph

4. _____ pseudopodium

5. _____ macrophage

a) A large cell that engulfs foreign material

b) An organism that gets energy from the sun

c) A cellular extension ("false foot") of amoeboid cells

d) Inflammation of the joints

e) Bacteria present in the blood

Answers: 1. d 2. e 3. b 4. c 5. a

Here is another rule to remember—spelling counts! In some cases, changing a single letter can make a difference in the meaning of the word. For example, *ilium* and *ileum*. The ilium is part of your hip bone, but the ileum is the last part of your small intestine. Something that is *intercellular* goes between cells, while something that is *intracellular* stays within a single cell.

Realize the importance of spelling—changing even one letter may change what you are referring to. ■

Another maneuver that can be tricky with these terms is converting them from the singular form to the plural form. Some of the rules are the same as in English. If you look at one mushroom and then another mushroom, you are indeed examining two mushroom**s**, and more than one bone are bone**s**. But there are some unique rules for pluralizing many terms that you will encounter. More than one vertebra are verte-bra**e**, and the small bone at the end of each of your fingers is a phalanx, but all of these bones in your fingers together are called phalan**ges**. **Table 3.2** shows the rules to guide you with pluralization.

TABLE 3.2   Basic rules for changing words in singular form to plural.

If the word ends in	Do this first	Then add	Examples
-a		Add -e	*alga* becomes *algae*
-ax	Drop -ax	Add -aces	*thorax* becomes *thoraces*
-ex or -ix	Drop -ex or -ix	Add -ices	*cortex* becomes *cortices*
-ma		Add -ta	*stoma* becomes *stomata*
-is	Drop -is	Add -es	*diagnosis* becomes *diagnoses*
-nx	Change -x to -g	Add -es	*larynx* becomes *larynges*
-on	Drop -on	Add -ia	*mitochondrion* becomes *mitochondria*
-us	Drop -us	Add -i	*nucleus* becomes *nuclei*
-um	Drop -um	Add -a	*bacterium* becomes *bacteria*
-y	Drop -y	Add -ies	*biopsy* becomes *biopsies*

In this chapter, we threw a lot of new terms at you. Most chapters in your textbook will do the same. At times, you may think that learning all the terms is impossible. Just when you think you know the name of a structure, you might learn that it has another name as well. Think about the red blood cell that carries oxygen in your blood. Do you know any other names for a red blood cell? You might—it is also called an *erythrocyte,* or it goes by the abbreviation RBC.

Multiple names are common in biology. At times, the confusion may seem intentional, but realize that there are many ways to name things—formal names (erythrocyte) and informal names (red blood cell). Mastering the language can be challenging, but you are now armed with important tools to help you. You will also find that many terms are defined in your textbook, either in the chapters or in the glossary. And you can always purchase a biological dictionary or refer to an online version—they are invaluable.

You now know to start your journey through each chapter in your textbook by learning the language. Know the terms so you can understand the material. Don't just memorize words—look up their meanings, look at how they are built, make and use flash cards, and be able to use your new words. Get in the habit of talking biology with your study group and with other people in your life, if possible. Learn what you can now, and the rest will come as you go.

People often move to foreign-speaking countries and learn the language as they go. They learn by experience and practice. Students have been surviving in the foreign-sounding world of microbiology for ages as well, and you have already had a crash course in the language. Enjoy your adventure!

Final Stretch!

Now that you have finished reading this chapter, it is time to stretch your brain a bit and check how much you learned.

WHAT DID YOU LEARN?

Try these exercises from memory first; then go back and check your answers, looking up any items that you want to review. Answers to these questions are at the end of the book.

PART A: USING TABLE 3.1, MATCH THE FOLLOWING TERMS WITH THEIR DESCRIPTIONS.

1. _____ leukocyte

2. _____ epitope

3. _____ phototaxis

4. _____ antihistamine

5. _____ cytoplasm

6. _____ hydrolysis

7. _____ photosynthesis

8. _____ phycology

9. _____ gastroenteritis

10. _____ hemagglutinate

a) To cause blood cells to clump

b) The process of converting light energy to chemical energy

c) Inflammation of the stomach and intestine

d) The study of algae

e) A white blood cell

f) A region on the surface of an antigen

g) The fluid and structures inside a cell

h) Breaking apart a compound by adding water

i) Movement in response to light

j) A drug that neutralizes histamine

PART B: ANSWER THESE QUESTIONS.

1. Which of the following is an eponym?

   _____

   a) MHC   b) MRSA
   c) steptococcus   d) Burkitt's lymphoma

2. Which of the following is an acronym?

   _____

   a) CDC   b) WHO
   c) staphylococcus   d) Hansen's disease

3. Using Table 3.1, construct a descriptive term for each of the following:

   Splitting of a glucose molecule _____

   _____

   Process of moving materials out of a cell

   _____

   An organism that likes just a small amount of air _____

   _____

4. Using the same table, define each of the following terms:

   Cytokinesis _____

   _____

   Chromosome _____

   _____

   Histopathology _____

   _____

5. Provide the plural form of the following terms:

   pharynx _____

   mitochondrion (structure inside a cell)

   _____

   flagellum _____

PART C: WRITE THE CHAPTER TERMS IN A NOTEBOOK AND DEFINE THEM IN YOUR OWN WORDS. GO BACK THROUGH THE CHAPTER TO CHECK YOUR MEANINGS, CORRECTING THEM AS NEEDED. LIST EXAMPLES WHEN APPROPRITE.

Descriptive term   Prefix
Eponym   Root
Abbreviation   Suffix
Acronym

4 Chemistry

The Science of Stuff

When you complete this chapter, you should be able to

■ Describe the hierarchy of organization of living things.

■ Compare and contrast the different states of matter.

■ Describe atomic structure.

■ Read and understand the periodic table of the elements.

■ Explain ionic, covalent, and hydrogen bonding.

■ Discuss basic types of reactions.

Your Starting Point

Answer the following questions to assess your chemistry knowledge.

1. The most basic unit of a chemical substance is the _____

2. What are the three states of matter? _____

   _____

3. An atom is made of what three subatomic particles? _____

   _____

4. What is the molecular formula for water? _____

5. What are three common types of chemical bonds? _____

   _____

6. What happens in anabolic reactions? _____

   _____

7. What is meant by an "organic" molecule? _____

   _____

Answers: 1. atom 2. solid, liquid, and gas 3. protons, neutrons, and electrons. 4. H_2O 5. ionic, covalent, and hydrogen bonds 6. Larger molecules are made from smaller molecules. 7. It contains both carbon (C) and hydrogen (H).

Climbing the Ladder: **The Biological Hierarchy of Organization**

Now that you are armed with some new language skills, let's peek at what lies ahead in your study of microbiology. In science, we like to categorize. One classification system is known as the biological hierarchy of organization. This hierarchy begins at the simplest level of structural organization and ends at the most complex, as shown in **Figure 4.1**. These levels, from the simplest to the most complex, are the atom, molecule, macromolecule, organelle, cell, tissue, organ, organ system, organism, population, community, ecosystem, and biosphere.

The bottom three levels shown in Figure 4.1 are addressed in **chemistry**, the study of elementary forms of matter, which we will review in this chapter. The next two levels are studied in **cytology** or cell biology, which we will explore in Chapter 6. **Organelles** are complex structures made of macromolecules. They carry out specific functions. **Cells** are the basic unit of living organisms. These first five levels are very important in microbiology.

The next, more complex levels—tissue, organ, and organ system—are the realm of human anatomy and physiology. **Tissues** are groups of cells organized to perform common functions. Tissues are organized into larger functional units called **organs**, such as the lungs, heart, and liver. Each organ performs at least one specific job. Multiple organs combine to form **organ systems**, each with some overall functions. For instance, the cardiovascular system pumps blood throughout the body.

The **organism** is the whole of these individual parts. Some organisms, such as bacteria, are merely a single cell. Other organisms are complex, consisting of billions of cells ultimately organized within complex organ systems.

Finally, the most complex levels—populations, communities, ecosystems, and the biosphere—involve more than one organism. A **population** is a localized group of organisms belonging to the same species; populations of different species living in the same area make up a **community**. Interactions between the living organisms in a community and the nonliving features of the environment (such as sunlight and water) form an **ecosystem**. The largest and most complex level of all is the **biosphere**, which encompasses the environments that are inhabited by life.

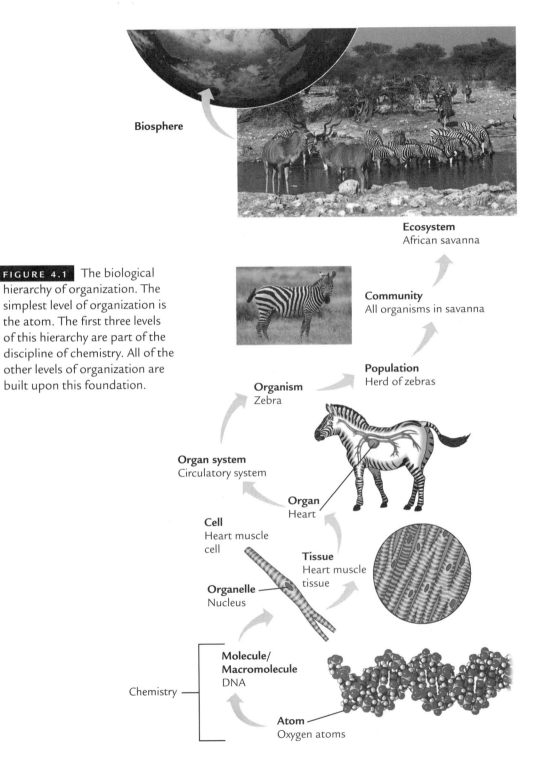

Biosphere

Ecosystem
African savanna

Community
All organisms in savanna

Population
Herd of zebras

FIGURE 4.1 The biological hierarchy of organization. The simplest level of organization is the atom. The first three levels of this hierarchy are part of the discipline of chemistry. All of the other levels of organization are built upon this foundation.

Organism
Zebra

Organ system
Circulatory system

Organ
Heart

Cell
Heart muscle cell

Tissue
Heart muscle tissue

Organelle
Nucleus

Chemistry

Molecule/ Macromolecule
DNA

Atom
Oxygen atoms

Beginning the Climb: **Chemistry Is Important**

Why do you have to learn chemistry before you learn biology and microbiology? Biology is the study of living organisms—including all of the "stuff" that they're made of. Chemistry is the science that covers all of that "stuff." In addition, all of the work done in cells involves chemical reactions. Chemistry is very much a part of our everyday lives.

In this chapter, we'll explore basic chemistry concepts to give you a head start for your current class. Learning chemistry may seem tough at times because the terminology can be challenging. For example, just read the ingredient list on almost any food product. Do we really eat all that stuff? But, as you learned in the last chapter, knowing the parts of words helps you to understand the meaning.

What's the **Matter?**

Let's start with something you already know about: **matter**. All the "stuff" of which you are made is matter, and matter is defined as anything that

■ Has mass (or weight), and

■ Takes up space

The terms *mass* and *weight* are often used interchangeably, but there is a difference. Mass refers to the actual physical amount of a substance. Weight takes into account the force of gravity acting on that mass. Consider astronauts. Each has a certain mass—the actual amount of material his or her body contains. Astronauts are weightless during space travel when there is no gravity, but their individual mass does not change. The second part of the definition of matter is that it takes up space. The space it occupies is called **volume**.

Matter typically exists in any of three physical states: solid, liquid, or gas. Let's consider water. What do we call the three states of water?

Solid: _____   Liquid: _____   Gas: _____

Those should be easy. Solid water is ice, liquid water is water (duh!), and its gas form is vapor, or steam. See—you already know chemistry!

Now, how can you change solid water to its gaseous form?

_____

When you add heat, which is a form of energy, ice melts to become a liquid. With enough heat, the liquid eventually boils to become vapor. If you collect the vapor and cool it, it will condense back to liquid. If you cool it enough, it will become ice (solid). As you can see, the three forms of matter are interchangeable.

✔ **QUICK CHECK**

What are the three states of matter? _____

Answer: solid, liquid, and gas

It's **Element**-ary, My Dear Watson!

All matter is composed of **elements**, which are the most basic chemical substances. More than 110 elements are recognized, and around 90 of these occur naturally on Earth. Some elements you probably know are iron, copper, silver, gold, aluminum, carbon, oxygen, nitrogen, and hydrogen. Some exist in pure form, such as helium and neon, but most occur in combination with other elements.

For the most part, living organisms require only about 20 elements. By weight, 95% of the human body is composed of just four of these:

- Carbon

- Hydrogen

- Oxygen

- Nitrogen

Each element is represented by a symbol, typically the first one or two letters of the element's name. The symbols for some of the elements, such as the four just listed, are quite logical. Others are less obvious. For example, the symbol for silver is Ag, but that is because it comes from the Latin word *argentum*, meaning *silver*.

TIME TO TRY

Several elements have names that begin with the letter C, so most of them use a two-letter chemical symbol. Try to match each of the following chemical names with their symbols.

Your choices	Names	Symbols
_____	calcium	Cu
_____	chromium	C
_____	cobalt	Ca
_____	copper (Latin = *cuprum*)	Co
_____	carbon	Cr

Answers: calcium = Ca, chromium = Cr, cobalt = Co, copper = Cu, carbon = C

Chemical Carpentry: **Atomic Structure**

All chemical elements are composed of tiny particles called **atoms**. An atom is the smallest complete unit of an element—one atom of carbon, for example, is the smallest piece of carbon that can exist. Two or more atoms can combine together to form larger structures called **molecules**. And simple molecules can join together to form more complex chemical structures called **macromolecules**, including proteins, carbohydrates, deoxyribonucleic acid (DNA), and fats. But they all begin the same way—with atoms.

Atoms vary in size, weight, and how they interact with other atoms, but they all share some common characteristics. All are made of smaller units called **subatomic particles** that are arranged in a precise manner. Although many subatomic particles are now recognized, the main ones of interest to us are protons, neutrons, and electrons.

THE NUCLEUS

The **nucleus** of an atom is not a structure. Instead, think of the nucleus as the area in the middle of an atom where some of the subatomic particles hang out. You should merely think of it as the atom's central region. Here we find two types of relatively large subatomic particles called **protons** and **neutrons.** Protons and neutrons have a similar size and about the same mass. Protons are positively charged particles and may be designated as p^+. Neutrons carry no electrical charge, and they are, as their name suggests, neutral. Neutrons may be designated by n^0, indicating they lack any electrical charge, or simply by **n**. All of the protons and neutrons in an atom are located in the nucleus.

ELECTRONS

The other major subatomic particles—the **electrons**—are in constant motion around the nucleus; they are never *in* the nucleus. Electrons are very small and have almost no weight. They carry a negative charge and are often designated as e^-. The number of negatively charged electrons orbiting the nucleus of an atom equals the number of protons in the nucleus. The negative charges of the electrons exactly balance the positive charges of the protons. Thus, any atom is, overall, neutral.

The number of e^- equals the number of p^+ in an atom. Thus, the atom is electrically neutral. ▪

Let's start with a simple image to get our bearings: the rings of the planet Saturn. The rings never touch the planet itself. These rings can represent the paths of the electrons around the nucleus. Unlike Saturn's rings, however, the electrons do not travel in a nice, even path along a single plane. Rather, they zip about quite rapidly in multiple paths called **orbitals** that have different shapes and orientations. A map of all the possible orbitals of an electron looks like a symmetrical cloud. This map, called an electron cloud, does not show actual electrons, but rather the probability of where the electrons are at any moment.

PICTURE THIS

You arrive home late one evening, well after the sun has set. Your front light is on. You glance overhead and see a large cloud of insects swarming around the light. This is the basic image you should have of the electrons orbiting the nucleus of an atom (**Figure 4.2**). They are constantly in motion around the nucleus, but their individual paths vary.

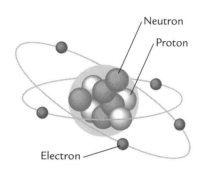

Neutron

Proton

Electron

FIGURE 4.2 Electrons orbiting around the nucleus of an atom of boron.

✔ QUICK CHECK

What are the three subatomic particles, and where is each located in an atom? _____

orbiting around the nucleus.

Answer: Protons and neutrons are always in the nucleus; electrons are always

ATOMIC NUMBER

All atoms of a given element have the same specific number of protons. Therefore, each element has its own **atomic number**, defined as the number of protons in an individual atom of that element. For example, hydrogen has an atomic number of 1. If you gave hydrogen another proton, it would no longer be hydrogen—it would now become a different element—helium, with an atomic number of 2. So, in order for atoms to be of the same element, they must all have the same number of protons.

The atomic number equals the number of protons in an atom. ■

Now, recall that an atom is usually electrically neutral. This means that the number of positive charges from protons must equal the number of negative charges from electrons. In other words, the number of protons in an atom always equals the number of electrons. Because of this, if you know an atom's atomic number, you know not only how many protons it has, but also how many electrons it has—they are the same number! Let's try this.

TIME TO TRY

Nitrogen's atomic number is 7.

1. How many protons does an atom of nitrogen have? _____

2. How many electrons does an atom of nitrogen have? _____

If nitrogen's atomic number is 7, that tells you it has seven protons. You know the number of protons must equal the number of electrons, so an atom of nitrogen will also have seven electrons. The seven positive charges from the protons are balanced by the seven negative charges from the electrons, so the atom is electrically neutral.

The number of protons equals the number of electrons in an atom. ■

ATOMIC MASS (WEIGHT)

An element's **atomic mass** refers to the total mass of a single atom of that element. Electrons are so tiny that they have almost no mass. Almost all of the atomic mass, then, comes from the combined masses of the larger protons and neutrons. But what do they weigh? Obviously atoms are too tiny for their mass to be measured in pounds or ounces, so imagine trying to weigh a subatomic particle. Conveniently, scientists developed a unit of measurement called the **atomic mass unit (u)**, and 1 u is approximately the mass of one proton. Neutrons have almost the same mass. The mass of an atom becomes amazingly simple! To determine an atom's mass, all you do is add its total number of protons and neutrons together. Because each of them weighs about 1 u, the atomic mass is approximately equal to the total number of protons and neutrons—and you can simply ignore those tiny little electrons.

Atomic mass equals the number of protons plus the number of neutrons in an atom. ■

For all atoms of an element, the number of protons is constant. But the number of neutrons can vary, so atoms of the same element can have different atomic masses. For example, iodine (atomic number 53) has 53 protons, 53 electrons, and *usually* 74 neutrons, so its atomic mass is approximately 127. But some atoms of iodine have 72 neutrons, giving an atomic mass of 125, and some have 78 neutrons, giving an atomic mass of 131. Atoms of the same element that have different atomic masses are called **isotopes**.

WHY SHOULD I CARE?

Some isotopes are radioactive, meaning they emit certain types of energy. The energy they emit can often be seen with special equipment. For this reason, some radioactive isotopes are used in medicine. For example, the thyroid gland uses iodine to make certain hormones. If a patient might have a thyroid problem, a radioactive isotope of iodine (^{131}I) can be injected into the blood. Then the clinician can use an imaging technique to monitor how well the thyroid is working. In another use, cobalt (^{60}Co) can be injected into an area where there is cancer, which irradiates and kills the tumor cells.

Is That an Eye Chart or a **Periodic Table of the Elements?**

Look at **Figure 4.3.** YIKES! It may look a bit scary at first, but that's only if you don't know how to read it. This is a partial **periodic table of the elements.** Let's explore how the table is arranged.

From the periodic table, we know that hydrogen's atomic number is 1, meaning it has one proton, which is shown in the center of the atom at the nucleus in **Figure 4.4a.** This means it also has one electron, which is shown orbiting the proton, in the first shell (the orbital around the nucleus). It rarely has any neutrons. Helium's atomic number is 2. The helium atom in **Figure 4.4b** has two protons and also two neutrons. (What is its atomic mass? _____) Helium also has two electrons, as shown. Next, look at lithium (**Figure 4.4c**), which has atomic number 3. You see it has three protons and three neutrons in its nucleus, and three electrons in orbit.

1A																	8A
1 **H** 1.008	2A											3A	4A	5A	6A	7A	2 **He** 4.003
3 **Li** 6.941	4 **Be** 9.012											5 **B** 10.81	6 **C** 12.01	7 **N** 14.01	8 **O** 16.00	9 **F** 19.00	10 **Ne** 20.18
11 **Na** 22.99	12 **Mg** 24.31	3B	4B	5B	6B	7B	⌐———8B———⌐		1B	2B		13 **Al** 26.98	14 **Si** 28.09	15 **P** 30.97	16 **S** 32.07	17 **Cl** 35.45	18 **Ar** 39.95
19 **K** 39.10	20 **Ca** 40.08	21 **Sc** 44.96	22 **Ti** 47.87	23 **V** 50.94	24 **Cr** 52.00	25 **Mn** 54.94	26 **Fe** 55.85	27 **Co** 58.93	28 **Ni** 58.69	29 **Cu** 63.55	30 **Zn** 65.41	31 **Ga** 69.72	32 **Ge** 72.64	33 **As** 74.92	34 **Se** 78.96	35 **Br** 79.90	36 **Kr** 83.80
37 **Rb** 85.47	38 **Sr** 87.62	39 **Y** 88.91	40 **Zr** 91.22	41 **Nb** 92.91	42 **Mo** 95.94	43 **Tc** (98)	44 **Ru** 101.1	45 **Rh** 102.9	46 **Pd** 106.4	47 **Ag** 107.9	48 **Cd** 112.4	49 **In** 114.8	50 **Sn** 118.7	51 **Sb** 121.8	52 **Te** 127.6	53 **I** 126.9	54 **Xe** 131.3
55 **Cs** 132.9	56 **Ba** 137.3	57 **La** 138.9	72 **Hf** 178.5	73 **Ta** 180.9	74 **W** 183.8	75 **Re** 186.2	76 **Os** 190.2	77 **Ir** 192.2	78 **Pt** 195.1	79 **Au** 197.0	80 **Hg** 200.6	81 **Tl** 204.4	82 **Pb** 207.2	83 **Bi** 209.0	84 **Po** (209)	85 **At** (210)	86 **Rn** (222)
87 **Fr** (223)	88 **Ra** (226)	89 **Ac** (227)	104 **Rf** (261)	105 **Db** (262)	106 **Sg** (266)	107 **Bh** (264)	108 **Hs** (269)	109 **Mt** (268)	110 **Ds** (271)	111 **Rg** (272)	112 — (285)	113 — (284)	114 — (289)	115 — (288)			

FIGURE 4.3   A partial periodic table of the elements.

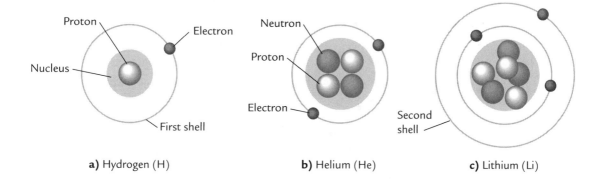

a) Hydrogen (H)   b) Helium (He)   c) Lithium (Li)

FIGURE 4.4 Illustrations of atoms of **a)** hydrogen, **b)** helium, and **c)** lithium, showing the placement of the electrons around the nucleus. The first shell fills first and can hold only two electrons. The next shell holds a maximum of eight.

TIME TO TRY

In the following illustrations, assume that the gray sphere in the middle represents the nucleus with its neutrons and protons. Add the shells and electrons for each of the elements. Boron is done as an example. Don't worry about the positions of the electrons; just draw the right number of electrons on the right number of shells.

Boron
Atomic number 5

Oxygen
Atomic number 8

Neon
Atomic number 10

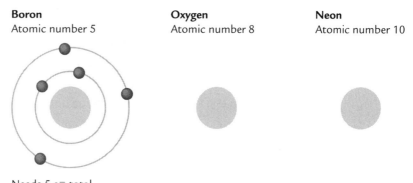

Needs 5 e⁻ total.
2 e⁻ are in the inner shell,
the other 3 in the outer shell.

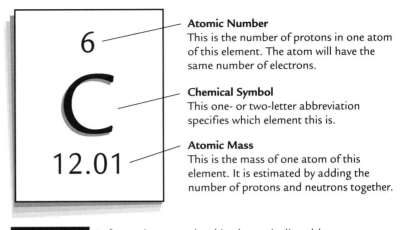

Atomic Number
This is the number of protons in one atom of this element. The atom will have the same number of electrons.

Chemical Symbol
This one- or two-letter abbreviation specifies which element this is.

Atomic Mass
This is the mass of one atom of this element. It is estimated by adding the number of protons and neutrons together.

FIGURE 4.5   Information contained in the periodic table.

In your drawings, oxygen should have eight electrons (two on the inner shell and six on the outer shell). Neon should have 10 electrons (two on the inner shell and eight on the outer shell).

Now look at **Figure 4.5**. This is the square from the periodic table that represents carbon. You can see that each square of the table tells you an element's chemical symbol, its atomic number, and its atomic mass. What do those three items tell you? _____

_____

Remember, if you know an element's atomic number, you know how many protons are in its atoms. Once you know that, you also know how many electrons it has because the number of protons and electrons is the same. But look at the atomic mass. Carbon's atomic mass is 12.01. Atomic mass equals the number of protons plus the number of neutrons. If the atomic mass is not a whole number, then you can round up or down to a whole number to figure out the typical number of neutrons.

TIME TO TRY

Use the periodic table (Figure 4.3) to answer the following questions.

1. How many protons are there in an atom of phosphorus (P)? _____

2. How many electrons are there? _____

3. How many neutrons does a typical atom of phosphorus have?

_____

You should see that the atomic number of phosphorus is 15, so it has 15 protons and therefore also 15 electrons. Its average atomic mass is 30.97, which rounds up to 31. So,

$$31 - 15 \text{ protons} = 16 \text{ neutrons}$$

See—this is just simple math!

Bumper Cars and **Chemical Interactions**

Have you ever tried bumper cars? If not, you really should—it's a great way to release tension. Atoms interact with each other rather like bumper cars do (**Figure 4.6**). The first part of a bumper car that makes contact with another car is the outer rubber bumper. When two atoms come together, the first parts to make contact are always the electrons in the outer shells. The protons and neutrons are safely tucked away in the nucleus at the center of the atom. So, the electrons in the outermost shell act as the "bumper" and determine how atoms interact with each other. Because all atoms of a particular element have the same number of electrons, all atoms of a particular element will react the same way.

In bumper cars, the outer rubber bumpers of the cars make contact first. The riders inside the car never contact each other.

In atoms, the outer shell electrons make contact first. The protons and neutrons inside the nucleus never contact each other.

FIGURE 4.6   Atoms interact somewhat like bumper cars. The electrons in the outer shells become the atoms' "bumpers" and determine the chemical reactivity.

TIME TO TRY

Consider carbon in the periodic table.

What is its atomic number? _____

How many protons does it have? _____

How many electrons? _____

Once you see that carbon's atomic number is 6, you know that it has six protons, and so it also has six electrons. Now draw the electrons for carbon around its nucleus.

There are two electrons on the inner shell and four electrons on the outer shell. These outermost electrons will interact with other electrons on the other atoms.

    The number of electrons in an atom determines its chemical reactivity. ▇

The Union: **Chemical Bonding**

As mentioned earlier, two or more atoms can join together to form a molecule. An atom's electrons orbit around the nucleus in one or more "shells." Each shell holds a specific maximum number of electrons. As you learned in Figure 4.4, the first shell (closest to the nucleus) holds only two electrons. The next shell holds eight; the third shell holds eight if it is the outermost, or 18 if it isn't. The outermost shell is called the **valency shell**. It can contain at most two electrons for helium or eight electrons for all other elements. If this outermost shell contains the maximum number of electrons, or is full, the atom is amazingly stable.

It is said to be chemically **inert**—it will not easily react with other atoms. It is stable and does not want to change. All of the elements in the last column of the periodic table are inert.

On the other hand, atoms of elements in all of the other columns lack a full outermost shell. That means they are unstable and want to become stable. Unstable atoms can gain, lose, or share electrons with other unstable atoms until they become stable. That's how atoms interact. Let's explore this more deeply.

✔ **QUICK CHECK**

Under what circumstances is an atom stable? _____

Answer: An atom is stable when the outermost, or valency, shell is full, meaning it has two electrons for helium or eight electrons for all other elements.

IONIC BONDING

When atoms become stable by losing or gaining electrons, electrons actually leave one atom's outermost shell and join the outermost shell of another atom. The atoms are now stable, meaning they each have a full outermost shell. However, gaining or losing electrons also changes the atoms in another way. Recall that atoms are normally electrically neutral—they have the same number of protons (+) and electrons (−). Once the electrons move, though, the atoms are no longer neutral because the number of protons and electrons are not equal. An atom that gains an electron has one extra negative charge, and an atom that loses an electron is short one negative charge, making it positive.

Atoms that have gained or lost electrons carry an electrical charge and are called **ions**. These are designated with a + or − sign. For example, sodium tends to lose an electron and become a sodium ion, **Na+**. Chlorine tends to gain an electron, becoming a chloride ion, **Cl−**. Ions of opposite charges attract each other ("opposites attract"). Whenever oppositely charged ions join, they form an **ionic bond**—ions form ionic bonds.

Atoms that gain or lose electrons form ions, and ions of opposite charges form ionic bonds. ▪

TIME TO TRY

Let's see how ionic bonding works, using sodium and chlorine. Look at the periodic table and fill in the following information:

	Sodium (Na)	Chlorine (Cl)
Atomic number:	_____	_____
Number of protons:	_____	_____
Number of electrons:	_____	_____

Draw the electrons around the nuclei of each of these atoms (the shells are drawn for you):

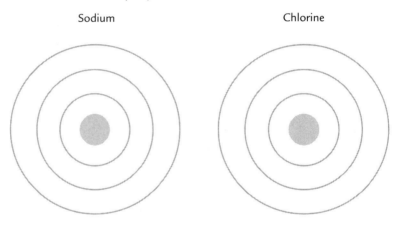

Sodium   Chlorine

Sodium, with atomic number 11, should have two electrons in its inner shell, eight in the second, and a single electron in its outermost shell. Chlorine, atomic number 17, should have two electrons in its inner shell, eight in its second shell, and seven in its outermost shell. That is convenient—sodium's valency shell has one too many, and chlorine's valency shell is short one. Sodium will lose its electron to chlorine, producing two ions, Na^+ and Cl^-. Once the ions are formed, their opposite electrical charges draw them together and they form a strong ionic bond, creating a molecule of a substance called *sodium chloride*. You know it better as *table salt*.

Now we can learn something else from an element's position on the periodic table: Sodium is in the first column, so it has one extra electron it wants to lose. Chlorine is in the next-to-last column, so it needs only one electron to have a full outermost shell and be stable.

What would you predict about calcium? _____

What about oxygen? _____

Calcium is in the second column, so it has two electrons in an outermost shell that wants eight. It is not stable. Oxygen is two columns short of being stable, so it needs two more electrons to fill its outermost shell and become stable.

For review, **Figure 4.7** shows how an ionic bond forms between lithium and fluorine. These elements are in the same columns as sodium and chlorine, so the process is the same except these new elements only have two shells of electrons. But, as you now know, in chemical interactions, only the electrons in the outermost shells are important.

✔ **QUICK CHECK**

How does an ionic bond form? _____

Answer: An ionic bond forms when atoms gain or lose electrons, forming oppositely charged ions that are then drawn together by their charges.

COVALENT BONDING

If you grew up with brothers or sisters, there were probably times when you disagreed. These disagreements may have revolved around possession of some toy or other item. You know how these squabbles usually ended—with a parental voice from somewhere in the distance yelling, "*share!*"

Some atoms are more "content" to share than to donate or receive. Let's consider two hydrogen atoms, each of which has a single electron. To form an ionic bond, one hydrogen atom would have to give up its electron, and another atom would have to gain it. Considering they're

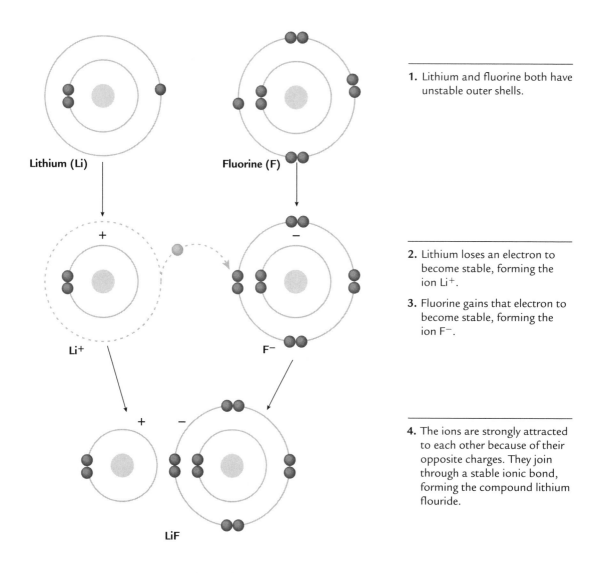

Lithium (Li)

Fluorine (F)

Li⁺

F⁻

LiF

1. Lithium and fluorine both have unstable outer shells.

2. Lithium loses an electron to become stable, forming the ion Li⁺.

3. Fluorine gains that electron to become stable, forming the ion F⁻.

4. The ions are strongly attracted to each other because of their opposite charges. They join through a stable ionic bond, forming the compound lithium flouride.

FIGURE 4.7   Ionic bonding. Lithium will lose an electron to fluorine, forming two oppositely charged ions. These ions are then attracted to each other and form an ionic bond, producing the compound called *lithium fluoride.*

identical, it's hard to determine which would gain and which would lose. Instead, both hydrogen atoms can share their two electrons. Both electrons orbit around both nuclei together—as a pair. Both atoms will be stable as long as they stay together because together they have a valency shell with two electrons, which is a full valency shell. This type of bond is called a **covalent bond** (see **Figure 4.8**). In general, elements that are closer to the right or left side of the periodic table are more likely to form ionic bonds, and those closer to the middle of the table are more likely to form covalent bonds.

✔ **QUICK CHECK**

How does a covalent bond form? _____

_____

Answer: A covalent bond forms when atoms share electrons to attain full outermost shells and become stable.

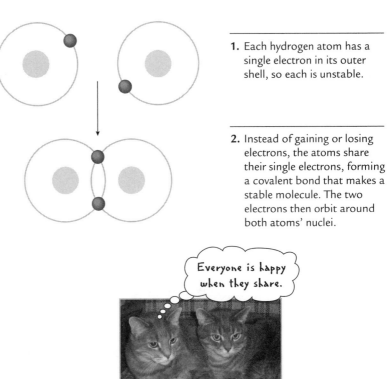

1. Each hydrogen atom has a single electron in its outer shell, so each is unstable.

2. Instead of gaining or losing electrons, the atoms share their single electrons, forming a covalent bond that makes a stable molecule. The two electrons then orbit around both atoms' nuclei.

Everyone is happy when they share.

FIGURE 4.8 Covalent bonding. In covalent bonding, instead of engaging in a tug-of-war, two atoms share the electrons in their outer shells to become stable. Here a covalent bond between two hydrogen atoms is shown.

HYDROGEN BONDING

Although there are many types of chemical bonds, we will look at just one more. A **hydrogen bond** is a weak bond that can form between the hydrogen atom of one molecule and a larger atom of another molecule. To understand this type of bond better, let's first look at a water molecule (**Figure 4.9**). A water molecule has two hydrogen atoms and one oxygen atom. A hydrogen atom is tiny, with just one electron and one proton. An atom of oxygen is much bigger—it has eight electrons, eight protons, and eight neutrons. Each hydrogen atom binds to oxygen by

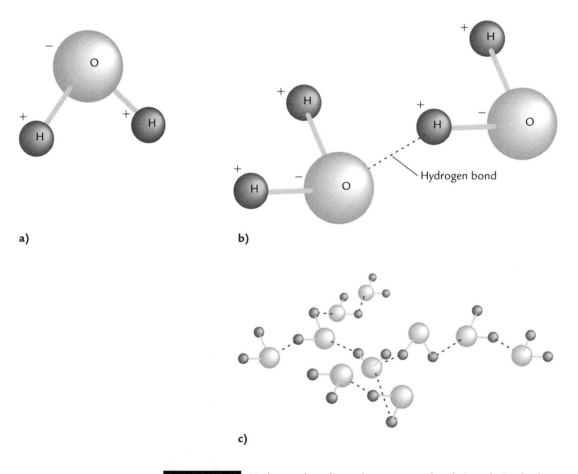

a)

b)

c)

FIGURE 4.9 Hydrogen bonding. **a)** A water molecule is polarized—the hydrogen atoms tend to be slightly more positive than that of the oxygen. **b)** A hydrogen bond forms between a slightly positive hydrogen atom and a slightly negative oxygen atom. **c)** Hydrogen bonds between water molecules give water many unique properties.

sharing its lone electron. Then the electron from hydrogen has to orbit around both nuclei—its own and also that of oxygen. The electrons from both hydrogen atoms spend more time around the oxygen than around their hydrogen because of the size difference. The greater number of protons in the nucleus of oxygen makes it more positive and, therefore, more attractive to the electrons than to the nucleus of hydrogen, which contains only one proton.

PICTURE THIS

Prove this to yourself with **Figure 4.9a**. Using your finger to represent the path of the shared electron, move your finger at a steady pace to trace the path around one hydrogen atom and move from there to pass around the oxygen atom. Your finger is near the oxygen longer.

In addition, oxygen only shares two of its electrons, so it always has six around it, even without the shared electrons. So at any given time, there are more electrons around the oxygen than around the hydrogen. As a result, there is a slight imbalance in the electrical charge of the water molecule—the oxygen tends to be a bit more negative than the hydrogens because the electrons hang out there more. This slight charge imbalance is called **polarity**.

Polar molecules can form weak hydrogen bonds with other molecules, as shown in **Figure 4.9b**. The slightly more positive hydrogen is attracted to the slightly more negative oxygen in an adjacent water molecule, so they form a weak bond. This is the hydrogen bond. Although they are weak, hydrogen bonds are important. They help shape many important molecules, such as proteins, your individual hairs (straight or curly), and your DNA.

Water has lots of hydrogen bonds, as shown in **Figure 4.9c**, giving it many unique characteristics. For example, ice floats because the hydrogen bonds cause the water molecules to spread out more in the solid form than they do in the liquid form. Water also has a high boiling point because of its hydrogen bonds. In addition, water has a high *adhesion*, meaning it sticks to surfaces well, and high *cohesion*, meaning its molecules stick to each other.

TIME TO TRY

Let's examine the compatibility of polar and nonpolar substances.

1. Take any clear container, preferably one that can be closed. A plastic baggie will work. Fill it about a third full with water.

2. Add to that about half as much cooking oil. Try to get them to mix and observe what happens. _____

_____

3. Add a few drops of food coloring to the container and shake it well to mix.

4. Is the food coloring polar or nonpolar (*Hint: Into which layer does it settle?*) _____

Water readily dissolves polar molecules, but it causes nonpolar molecules to bunch together.

✔ **QUICK CHECK**

Explain what is meant by *polar molecule*. _____

Answer: A polar molecule is one in which there is an uneven charge distribution across the molecule, resulting in slightly positive and slightly negative charges.

MOLECULES AND COMPOUNDS

As mentioned earlier, when two or more atoms bind together, they form a molecule. If the atoms are from the same element, they form a molecule of that element; for example, O_2 is a molecule of oxygen. If the atoms are from different elements, the substance formed is called a **compound**. Water is a compound because it has two elements.

A molecule is described by a **molecular formula** that tells you what the molecule is made of. This formula includes the letter symbols for the elements and the number of atoms of each element that are present in the molecule. Here are some examples:

- H_2O = water

- O_2 = oxygen molecule

- CO_2 = carbon dioxide

- CO = carbon monoxide

Note that the only difference between carbon dioxide and carbon monoxide is one atom of oxygen. We make carbon dioxide in our body and exhale it with every breath, whereas carbon monoxide is a deadly poison.

The molecular formula gives us limited information. It tells us how many pieces are in a molecule but not how they are hooked together. For that, we consult the **structural formula**, which is a simplified drawing of how the molecule is built. Lines in a structural formula represent chemical bonds (see Figure 4.9). Now look at **Figure 4.10**. This shows the structural formulae for three sugars: glucose, galactose, and fructose. Glucose and galactose are similar, so the differences are highlighted. Fructose, also called fruit sugar, has an obviously different appearance.

TIME TO TRY

Look carefully at Figure 4.10 and fill in the following information.

	Glucose	Galactose	Fructose
Number of carbon atoms:	_____	_____	_____
Number of hydrogen atoms:	_____	_____	_____
Number of oxygen atoms:	_____	_____	_____
Molecular formula:	C_H_O_	C_H_O_	C_H_O_

(*Hint: Use the numbers you wrote for each element above.*)

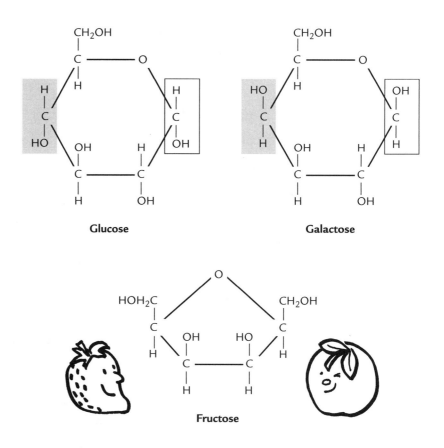

FIGURE 4.10 The structural formulae for glucose, galactose, and fructose. Although they all have the same molecular formula, the highlighted areas show the differences between glucose and galactose, and fructose is even more obviously different.

You can see that all three of these sugars have the same molecular formula, $C_6H_{12}O_6$. The structural formula provides more detailed information and is often more useful than the molecular formula, but the molecular formula is the most common method of describing molecules and compounds.

✔ QUICK CHECK

1. What is the basic difference between a molecule and a compound?

_____

_____

2. What is the difference between a molecular formula and a structural formula?   _____

_____

Answers: A molecule is formed whenever two or more atoms join. A compound is a molecule formed from two or more different atoms. 2. A molecular formula tells us how many atoms of each element are in a molecule. The structural formula gives us that information plus how those atoms are connected.

Is It **Organic?**

The term *organic* is used today to describe artwork, home décor, and how food is grown, among other things. In chemistry, however, it has a precise meaning. By definition, **organic compounds** contain both carbon and hydrogen. They must have these two elements, and they usually have others as well. Most of your body is made of organic molecules. These are the main categories of organic compounds:

- Carbohydrates (sugars, glycogen, and starches)

- Proteins (enzymes, some hormones)

- Lipids (which include fats and steroid hormones)

- Nucleic acids (DNA, which is your genetic material, and ribonucleic acid [RNA], which assists DNA)

   Coincidentally, these categories of organic molecules are also considered **macromolecules** because they are larger molecules made up of smaller molecules. For instance, a protein is made of chains of smaller molecules called *amino acids*; a nucleic acid is made of chains of smaller molecules called *nucleotides*.

Any molecules that do not contain both carbon and hydrogen are called **inorganic compounds**. These include carbon dioxide, oxygen, many acids and bases, salts, and water, which is our most important nutrient.

✔ **QUICK CHECK**

What is the difference between organic and inorganic molecules?

_____

_____

Answer: Organic molecules contain both hydrogen and carbon, while inorganic molecules do not contain both.

Love Is Just a **Chemical Reaction**

All activities that occur within living organisms involve chemical reactions. During chemical reactions, bonds between molecules or atoms are formed or broken to produce new chemical molecules or to release ions. Energy is required to make bonds, and when those bonds are broken, energy is released.

Chemical reactions are written in the form of a **chemical equation**, but instead of using an equals sign, we use an arrow. Let's look at this example:

$$Na^+ + Cl^- \rightarrow NaCl$$

The substances to the left of the arrow (Na^+ and Cl^-) are the **reactants**—the things that react together. The item to the right of the arrow ($NaCl$) is the **product**—the end result of the reaction. The arrow means "produces." Many reactions are **reversible**, meaning they can go in either direction. You can combine the ions we just mentioned to make salt, but salt can also break down to produce the ions. Reversible reactions are often indicated with a special double-arrow symbol: $\rightleftharpoons$.

Many biological reactions are assisted by a molecule called an **enzyme.** An enzyme is a compound that helps to bring together reactants. It is not permanently changed by the chemical reaction—it only makes the reaction happen more easily. Think of an enzyme as a sort of matchmaker. Without the enzyme to bring together the molecules, they might never meet.

Let's examine three basic types of reactions: synthesis, decomposition, and exchange.

- **Synthesis reactions** are reactions that build. Another term for these is *anabolic reactions.* (You might have heard of athletes who use *anabolic* steroids—to *build* their muscles.) In synthesis reactions, small atoms or molecules are combined to make a larger molecule. For example, small sugars combine to build large molecules of starch. Energy is required to make the bonds that hold together the larger molecules. Synthesis reactions are important in microorganisms for growth and repair processes.

- **Decomposition reactions** are the opposite of synthesis reactions. In this case, larger molecules are broken down into smaller molecules or atoms. These are also called **catabolic reactions.** For example, starch is broken down into smaller sugars. Salt is broken down to ions. Microorganisms use decomposition to digest food and gain energy from breaking the bonds.

- **Exchange reactions** involve swapping pieces. Two or more molecules split apart and then recombine in a new way; for example, AB and CD separate and then recombine to form AC and BD. Exchange reactions allow a cell to receive and store molecules in one form and then reuse them for multiple purposes.

WHY SHOULD I CARE?

All of the work done by the cells of living organisms—collectively referred to as **metabolism**—involves chemical reactions. We eat and breathe to bring in the necessary molecules. Then we use them in an amazing array of chemical reactions that allow us to do virtually everything we do. All body processes rely on synthesis, decomposition, and exchange reactions. When you begin studying specific chemical processes, it may help to understand them if you think of what the outcome should be—are the reactions building, breaking down, or swapping? If you know the goal, you can more easily understand the reactions.

✔ **QUICK CHECK**

What are the three basic types of chemical reactions? _____

_____

Answer: synthesis, decomposition, and exchange reactions

Atoms and molecules are the stuff of life, and the reactions between them are the processes of life. We hope now you can see that to achieve success in microbiology, chemistry truly does matter.

Final Stretch!

Now that you have finished reading this chapter, it is time to stretch your brain a bit and check how much you learned.

WHAT DID YOU LEARN?

PART A: WRITE THE CORRECT NUMBER IN THE BLANKS, USING THE PERIODIC TABLE IN FIGURE 4.3.

1. What is the atomic number of potassium? _____

2. What is the atomic mass of iodine? _____

3. For oxygen, write the number of protons _____ and electrons ____.

4. How many electrons are in the outer shell of iodine? _____

5. How many electrons can the first shell hold? _____ The second shell? _____

PART B: ANSWER THE FOLLOWING QUESTIONS.

1. Which subatomic particles are always in the nucleus? _____

   _____

2. Which subatomic particles are not included in the atomic mass? _____

   _____

3. A calcium ion has a +2 electrical charge. How did this ion form to give it that charge? _____

   _____

4. What is the basic difference between an ionic and covalent bond? _____

   _____

   _____

5. Place a check mark next to all of the compounds below that are organic.

   ____ $C_6H_{12}O_6$ _____ CO_2 _____ CH_2OH

   ____ NH_3 ____ HCl

6. If you get a cut on your finger, what type of chemical reaction will allow your skin to repair itself? _____

   _____

   _____

PART C: WRITE THE CHAPTER TERMS IN A NOTEBOOK AND DEFINE THEM IN YOUR OWN WORDS. GO BACK THROUGH THE CHAPTER TO CHECK YOUR MEANINGS, CORRECTING THEM AS NEEDED. LIST EXAMPLES WHEN APPROPRIATE.

Chemistry

Cytology

Organelles

Cells

Tissues

Organs

Organ systems

Organism

Population

Community

Ecosystem

Biosphere

Matter

Mass

Weight

Volume

Elements

Atoms

Molecules

Macromolecules

Subatomic particles

Nucleus

Protons (p^+)

Neutrons (n^0)

Electrons (e^-)

Orbitals

Atomic number

Atomic mass

Atomic mass unit (u)

Isotopes

Periodic table of the elements

Valency shell

Inert

Ions

Ionic bond

Covalent bond

Hydrogen bond

Polarity

Compound

Molecular formula

Structural formula

Organic compound

Inorganic compound

Chemical equation

Reactants

Product

Reversible

Enzyme

Synthesis (anabolic) reactions

Decomposition (catabolic) reactions

Exchange reactions

Metabolism

5 Biology Basics

How Life Works

When you complete this chapter, you should be able to

■ Describe the features common to all living organisms.

■ Discuss the major groups of living organisms.

■ Understand the basic process of scientific inquiry.

Your Starting Point

Answer the following questions to assess your knowledge.

1. What are the main features found in all living organisms?

_____

_____

2. What is evolution? _____

_____

3. Name three domains of living organisms.

_____

_____

Thinking about Life: **Basic Biological Principles**

Before we climb further up the ladder of biological organization, we need to summarize some of the principles common to all living things. These principles will help you understand why living organisms do what they do. They will also help you describe differences between living and nonliving things. These concepts should help guide your thinking as you try to understand the "whys" behind your learning.

Answers: 1. Living organisms are composed of cells, are able to maintain homeostasis, can reproduce on their own, will interact with and adapt to their environment, and can convert energy into usable forms. 2. Evolution is a generation-to-generation change in the proportion of different inherited genes in a population. 3. Three domains of living organisms are Archaea, Bacteria, and Eukarya.

TIME TO TRY

Place a checkmark next to the items you think are living.

___ Chair   ___ Milk   ___ Bacteria

___ Blade of grass ___ Bread mold   ___ Coral reef

___ Telephone   ___ Sugar   ___ Mushroom

LIFE BEGINS AT THE CELL

The **cell** has a unique place in the biological hierarchy of organization. It is the lowest level of structure that has the ability to perform all of the activities necessary for life, including that of reproduction. All cells share at least three common characteristics:

1. They are enclosed by a cytoplasmic membrane, which separates their internal environment from the external environment.

2. They are filled with cytoplasm, which is a mixture of substances in a liquid.

3. They contain DNA, which is the cell's genetic material.

TIME TO TRY

Examine the three drawings below. Circle the ones that you think are cells.

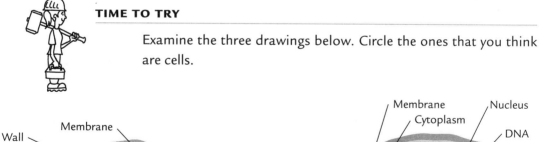

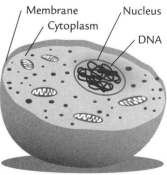

Cells have a cytoplasmic membrane, cytoplasm, and DNA. While all of the pictures have DNA, only the left and right diagrams represent cells. A **virus** (center diagram) lacks a cytoplasmic membrane and cytoplasm. Thus, even though they may have DNA (or RNA), viruses are not cells. Most microbiology texts place them in a unique category of nonliving forms. This category also includes prions and viroids, other entities which lack cellular structure. We will learn a little more about these unique, nonliving entities in Chapter 7.

All living organisms are made of cells. ▪

REGULATION AND HOMEOSTASIS

To maintain the utmost efficiency, living organisms exist in an optimal working environment. To achieve this, they have numerous regulatory mechanisms that work to maintain a relatively constant internal environment. This internal constancy is called **homeostasis**. You can think of it as maintaining the right balance of conditions within an organism.

TIME TO TRY

Turn back to Table 3.1 to determine the literal meaning of the word *homeostasis*. _____

Homeostasis literally means "to stay the same." Most work done within an organism is the result of chemical reactions. These reactions occur most efficiently if there is a relatively constant temperature, the right amount of water, and the right number of molecules.

We have an acceptable normal range of values for each of these in the human body. We know normal human internal body temperature is about 37° Celsius (98.6°F). If our body cools too much, chemical reactions occur more slowly or not at all. If our temperature rises, chemical reactions speed up and some important molecules may be destroyed. Microbial cells are also sensitive to temperature variations.

Water is the main component of cells. Water balance is absolutely critical to cells. If cells have too little water, nutrients cannot be adequately transported and wastes can accumulate to toxic levels. Cells would work less efficiently and may die.

WHY SHOULD I CARE?

Here is an example of how homeostasis is related to microbiology. A bacterium called *Salmonella* causes diarrhea, resulting in a significant loss of fluids from the human body. If not treated promptly, this infection can cause significant damage to the body's organs because of improper water balance.

REALITY CHECK

If *Salmonella* reproduces best at 37°C (or 98.6°F), how could your own body adjust to reduce the rate of *Salmonella* reproduction?

_____

_____

Many bacterial infections result in fever, an increase in body temperature above normal. This results in slower reproduction of bacteria and allows the body's defense mechanisms to work more efficiently.

REPRODUCTION AND INHERITANCE

It is only through reproduction that organisms pass on their traits and a species survives. In fact, many biologists maintain that the ultimate purpose of our existence is reproduction.

Tied to reproduction is inheritance, the sum of the genetic characters that are passed from parents to offspring. These characteristics are passed on to the next generation in the form of genes. **Genes** are made of the molecule **DNA,** which stores information. Most human cells (except sperm, ova, and red blood cells) contain a copy of all the DNA that you inherited from your mother and father. Before a cell divides, it copies its DNA and passes this genetic information on to each of the two cells it produces.

How is this information passed from parent organisms to offspring? In humans, it occurs during sexual reproduction, when an egg cell from the mother fuses with a sperm cell from the father. The result is a fertilized cell containing a combination of DNA from both parents. As you study microbiology, you will learn that microorganisms also pass on their unique characteristics in the form of DNA.

Reproduction enables survival of the species. During reproduction, genes, made of DNA, are passed from parent to offspring in the process of inheritance. ■

✔ **QUICK CHECK**

1. What is homeostasis? _____

_____

_____

2. What is the primary goal of any organism? _____

_____

Answers: 1. Homeostasis is a relatively constant internal environment.
2. reproduction

INTERACTION WITH THE ENVIRONMENT

Life doesn't exist in a vacuum. Each organism interacts continuously with its environment, which includes other organisms as well as nonliving components of the environment. The roots of a tree, for example, absorb water and minerals from the soil. Leaves take in a gas called carbon dioxide from the air. The tree also releases oxygen to the air, and its roots help form soil by breaking up rocks. Both organisms and their environments are affected by the interactions between them. The tree also interacts with other living things, including microscopic organisms in the soil that are associated with the plant's roots and animals that eat its leaves and fruit. We are, of course, among those animals. **Ecology** is the branch of biology that investigates these relationships between organisms and their environments, at the level of the ecosystem.

LIFE REQUIRES ENERGY

All living organisms have the independent ability to convert energy into a usable form. **Energy** is defined as the ability to do work. We can look at energy use on at least two levels: uses by the cell and uses by organisms in the ecosystem. Energy is used by all cells, allowing them to move, grow, reproduce, and perform all the other activities required for life.

Together, all of the chemical reactions that take place in a cell are called its **metabolism**. Plants, algae, and some bacteria harness the power of the sun by using a process called **photosynthesis,** in which solar energy is converted to the chemical energy contained in sugars. Such organisms are the **producers** in an ecosystem. (**Figure 5.1** depicts this energy flow through an ecosystem.)

Most organisms convert chemical energy from the sugars produced by photosynthetic organisms into a more usable form of energy—a molecule called adenosine triphosphate (ATP). **Consumers** are organisms such as animals, that feed on plants either directly (by eating plants) or indirectly (by eating animals that eat plants).

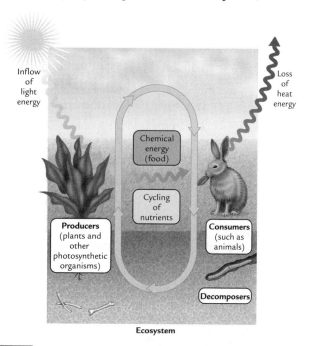

Inflow of light energy

Loss of heat energy

Chemical energy (food)

Cycling of nutrients

Producers (plants and other photosynthetic organisms)

Consumers (such as animals)

Decomposers

Ecosystem

FIGURE 5.1   Energy and nutrient flow in an ecosystem. Living is work, and work requires that organisms obtain and use energy. Most ecosystems are solar powered. The energy that enters an ecosystem as sunlight exits as heat, which all organisms dissipate to their surroundings whenever they perform work. In contrast, the nutrients within an ecosystem are recycled.

Thus, energy in an ecosystem flows from sunlight to producers and then to consumers. In the process some energy is lost as heat.

At both the cellular and ecosystem levels, energy is necessary for life. ■

Obtaining all the materials needed to make and maintain an organism's parts requires considerable energy. To minimize this cost, all organisms recycle. Many molecules are broken down to atoms and then built back into other molecules over and over again. Think about a child's building block set. Children can spend hours building things, taking them apart, and then building new ones—a car is stripped down, then becomes a castle, which is stripped down and rebuilt into a robot. Like the child, all living organisms reuse both energy and materials. Organisms that recycle chemicals in an ecosystem are known as **decomposers**.

All living organisms require energy to do work. ■

ADAPTATION AND EVOLUTION

Can you find the animal in **Figure 5.2**? It is a species of insect called a mantid. The shape and color enable it to blend into its background. This camouflage makes the mantid less visible to animals that feed on insects. It also makes it less visible to the insects that the mantids consume!

The unique characteristics that camouflage each mantid species are examples of adaptations. An **adaptation** is an inherited trait that helps the organism's ability to survive and reproduce in its particular environment.

How do mantids and other organisms adapt to their environments? Part of the answer is the variation among individuals in a population. Just as you and your classmates are not exactly alike, individual organisms within a population also vary in some of their traits. These variations reflect each individual's particular combination of inherited

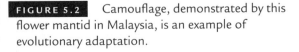

FIGURE 5.2   Camouflage, demonstrated by this flower mantid in Malaysia, is an example of evolutionary adaptation.

genes. This hereditary variation is the raw material that makes it possible for a population to adapt to its environment. If a particular variation is helpful, individuals with the variation may live longer and produce more offspring than those who do not have it. This process is called **natural selection** because the natural environment "selects" the individuals with certain inherited traits for survival.

Figure 5.3 illustrates a hypothetical example of natural selection in a bacterial population. A few of these bacteria have genes that allow them to be resistant to the antibiotic ampicillin. Now suppose that these bacteria are exposed to ampicillin. Ampicillin kills most of the susceptible bacteria, allowing the resistant bacteria more space to multiply and eventually become the most common type. This is the result of natural selection—the antibiotic "selected" for the resistant forms. Thus, the population has adapted to its environment.

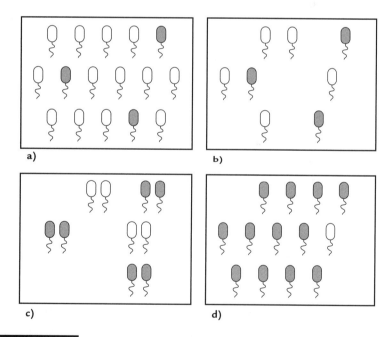

FIGURE 5.3   The original bacterial population **a)** consists of mostly cells that are susceptible to ampicillin. Only three of the bacteria (dark shading) contain genes to resist ampicillin. After ampicillin is added **b)**, some of the susceptible cells are killed immediately, while others are killed after additional time **c)**. In the meantime, the resistant cells have reproduced, and their offspring take the place of the killed cells **c)** and **d)**. Finally, most of the susceptible cells have been killed by the ampicillin, leaving resistant cells as the majority of the population **d)**. The population has adapted to ampicillin.

Natural selection is the mechanism by which evolution occurs. The word *evolution* means "a process of change." Biologists use the term **evolution** specifically to mean a generation-to-generation change in the proportion of different inherited genes in a population. In our bacterial example, genes for resistance to ampicillin are becoming more common, and cells that are susceptible are becoming less common in subsequent generations of bacteria. The bacterial population is said to be undergoing evolution, or evolving.

UNITY IN DIVERSITY

Diversity is a hallmark of life. Biologists so far have identified and named about 1.8 million species. This **biodiversity** includes more than 290,000 types of plants, 52,000 kinds of vertebrates (animals with backbones), and 1,000,000 kinds of insects. There are thought to be millions of bacterial species, many of which are not yet described. Estimates of the total diversity of species range from about 10 million to more than 200 million. Whatever the actual number, the vast variety of life widens biology's scope.

If life is so diverse, how can biology have unifying basic principles? What, for instance, can a tree, a bacterium, and a human possibly have in common? As it turns out, a great deal! Underlying the diversity of life is a striking unity, especially at the lower levels of the biological hierarchy of organization. You have already seen one example: the universal genetic language of DNA. That fundamental fact connects all of life. Another is the process that increases this amazing diversity: evolution by natural selection. Above the cellular level, however, organism adaptations vary so much according to their ways of life that describing and classifying biological diversity remains an important goal of biology.

WHY SHOULD I CARE?

Once we appreciate the diversity of microorganisms, we can then examine how those organisms respond to changes in climate (e.g., global warming) or other aspects of their environment (e.g., pollution). By studying the effects of these changes on microorganisms, we can predict the effect they might have on our own species.

Grouping Life: **Classification**

Biological diversity can be a bit overwhelming. Confronted with such complexity, there seems to be a human tendency to group diverse items according to similarities. For instance, perhaps you organize your music collection according to artist. And then maybe you group the various artists into broader categories, such as rock, jazz, country, or classical. In the same way, grouping similar species is natural for us. We may speak of squirrels and butterflies, though we recognize that many different species belong to each group. We may even sort groups into broader categories, such as rodents (which include squirrels) and insects (which include butterflies). **Taxonomy**, the branch of biology that names and classifies species, formalizes this ordering of species into a series of groups of increasing breadth. For now, we will focus on kingdoms and domains, the broadest units of classification.

Until recently, most biologists divided the diversity of life into five main kingdoms. (The most familiar are the plant and animal kingdoms.) But new methods, such as comparisons of DNA and RNA among organisms, have led to an ongoing reevaluation of old classification schemes. Biologists now group the kingdoms of life into three even higher levels of classification called domains. The three domains are named Bacteria, Archaea, and Eukarya (**Figure 5.4**).

The first two domains, **domain Bacteria** and **domain Archaea**, identify two different groups of organisms that have simple cells without a true nucleus or organelles. These are called **prokaryotes**. The cells of Bacteria and Archaea differ enough that these microorganisms occupy separate domains. In fact, Archaea are now thought to be more closely related to eukaryotes than to Bacteria.

All the **eukaryotes** (organisms made of cells with a true nucleus) can be grouped into the **domain Eukarya**. Historically, there were four kingdoms within the Eukarya:

■ Plantae. Plants are also multicellular organisms that are producers —they make their own food by using the energy of sunlight. Their cells are bound by a cytoplasmic membrane covered by a cell wall made of cellulose.

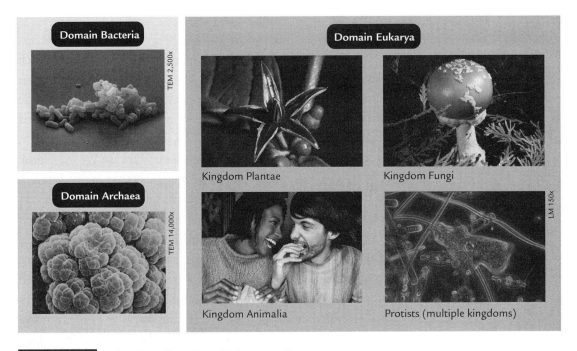

FIGURE 5.4   The three domains of living organisms.

- Animalia. Animals are multicellular organisms that are consumers—they obtain food by ingestion (eating) of other organisms. Their cells are bound only by cytoplasmic membranes and lack cell walls.

- Fungi (molds, mushrooms, yeasts). Fungi are also consumers. Many of them are decomposers that absorb nutrients by breaking down dead organisms and organic wastes, such as leaf litter and animal feces. Like plants, their cells are bound by both a cytoplasmic membrane and a wall, but their cell wall is made of chitin.

- Protista. Protists are single-celled organisms. However, their similarities stop there. Some, such as algae, are producers and have cell walls. Others, such as amoebae, are consumers and lack a cell wall. Therefore, most biologists now group these organisms into several smaller kingdoms.

✔ **QUICK CHECK**

Match the terms that apply to each kingdom of eukaryotes.

1. Plantae _____   a) Consumer
2. Fungi _____   b) Producer
3. Animalia _____   c) Has a cell wall

Answers: 1. b and c 2. a and c 3. a

How have scientists learned so much about living organisms? How can we be sure their discoveries are accurate? The next section will tell you more about how scientists do their work.

The Framework of Science: **The Scientific Method**

The word *science* is derived from a Latin verb meaning "to know." Science is a way of knowing. It developed from our curiosity about ourselves and the world around us. Scientists argued for hundreds of years whether living organisms could arise from nonliving things (spontaneous generation). Eventually French bacteriologist Louis Pasteur disproved spontaneous generation with a series of carefully planned experiments. Most bacteriologists accepted Pasteur's work because he constructed his experiments so that there could be no outside variables that influenced the results. Proper experimental design is part of a framework for the process of scientific research that is called the **scientific method.**

Figure 5.5 illustrates the steps in this framework. Observations (1) are made by one or more scientists. The study of these observations leads to one or more questions (2). The scientist then develops a hypothesis (3). The **hypothesis** is a proposed answer to the question and explains the observations. In order to test whether the hypothesis is correct, the scientist designs experiments (4) to test the hypothesis. Each experiment should be designed so that there is only one variable (cause) that could lead to the expected results. All other possible influences should be controlled so that they don't affect the results. The results of the experiments will cause the scientist to accept, modify, or

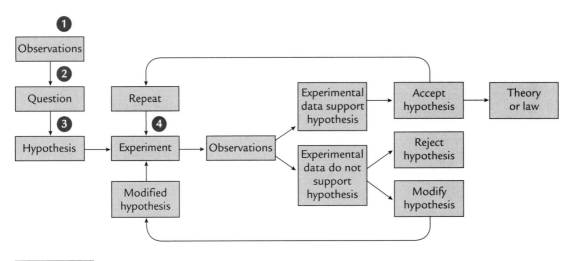

The scientific method, a framework for scientific research.

reject the hypothesis. If the same experiments are repeated by others and they always support the hypothesis, then it may become widely accepted as a theory or law. If, however, the experiments are repeated with different results or the results do not support the hypothesis, then the hypothesis is either modified or rejected and perhaps a new hypothesis is generated. Thus, the scientific method helps to ensure that results of an experiment are valid.

TIME TO TRY

Let's look at an example of how the scientific method might be used in microbiology. A microbiology student has collected and isolated a bacterial specimen from the floor of his microbiology lab classroom. He has observed that the bacterium grows well when he puts it in a tube of glucose solution and incubates it at 37°C. He wonders what the optimal growth temperature is for this bacterium. His hypothesis would be that 37°C is the optimal growth temperature. How should this student design an experiment to test his hypothesis? _____

_____

_____

You might have suggested that he try growing the bacterium at different temperatures and then visually examine the amount of growth to see where it grows best, but did you think about other factors that might influence those results? For instance, what if he added more bacteria to one tube than another? The amount of bacteria added would influence the amount of growth. Thus, he should find a way to add the same amount of bacteria to each tube. Should he use glucose solution to grow the bacterium, or could he use any available solution? The bacterium might grow better or not at all in a solution made of different substances. The amount of glucose in each tube should also be controlled. You can see from this example that experimental design must be carefully considered if the results and conclusions are to be valid.

You will undoubtedly use the scientific method when working in your microbiology laboratory. The same basic principles are used when medical personnel are trying to determine the cause of a new disease or test a new antibiotic to treat a disease.

✔ **QUICK CHECK**

What are the four basic steps of the scientific method? _____

_____

_____

Answers: 1. observations 2. question 3. hypothesis 4. experimentation

Final Stretch!

Now that you have finished reading this chapter, it is time to stretch your brain a bit and check how much you learned.

WHAT DID YOU LEARN?

Try these exercises from memory first; then go back and check your answers, looking up any items that you want to review. Answers to these questions are at the end of the book.

PART A: ANSWER THE FOLLOWING QUESTIONS.

1. List the features common to all living organisms. _____

_____

_____

2. What is homeostasis and why is it important? _____

_____

_____

3. What substance transmits information about characteristics from parents to offspring? _____

4. Where do most living things on Earth ultimately get their energy? _____

5. How are pesticide-resistant insects an example of natural selection in action?

_____

_____

6. Name the three domains and then name three kingdoms of eukaryotes. _____

_____

_____

_____

_____

7. What are the steps in the scientific method?

_____

_____

_____

_____

PART B: MATCH THE ORGANISMS ON THE LEFT WITH THEIR APPROPRIATE KINGDOM OR DOMAIN ON THE RIGHT.
(Hint: You may use a letter more than once.)

1. _____ butterfly

2. _____ mushroom

3. _____ amoeba

4. _____ fern

5. _____ algae

6. _____ grapevine

7. _____ bacterium

a) Domain Bacteria

b) Kingdom Animalia

c) Kingdom Fungi

d) Protists

e) Kingdom Plantae

PART C: WRITE THE CHAPTER TERMS IN A NOTEBOOK AND DEFINE THEM IN YOUR OWN WORDS. GO BACK THROUGH THE CHAPTER TO CHECK YOUR MEANINGS, CORRECTING THEM AS NEEDED. LIST EXAMPLES WHEN APPROPRIATE.

Cell
Virus
Homeostasis
Genes
DNA
Ecology
Energy
Metabolism

Photosynthesis
Producer
Consumer
Decomposer
Adaptation
Natural selection
Evolution
Biodiversity

Taxonomy
Domain Bacteria
Domain Archaea
Prokaryote
Eukaryote
Domain Eukarya
Scientific method
Hypothesis

6 Cell Biology

Life's Little Factories

When you complete this chapter, you should be able to

■ Discuss how organelles and cells fit into the biological hierarchy.

■ Describe the structure and function of the cytoplasmic membrane.

■ List the steps and structures in making a protein.

■ Describe the structure and function of the various eukaryotic cell organelles.

■ Describe the structure of a bacterial cell and explain how it differs from eukaryotic cells.

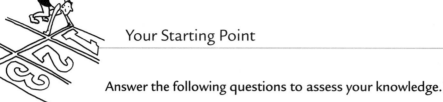

Your Starting Point

Answer the following questions to assess your knowledge.

1. What is an organelle? _____

_____

2. The main structural differences between animal and plant cells are

_____

_____

3. What is the function of a ribosome? _____

_____

4. Where is the DNA located in a eukaryotic cell? _____

_____

Where is the DNA located in a prokaryotic cell? _____

_____

5. Genes are made of what molecule? _____

Answers: 1. An organelle is a group of molecules and macromolecules that carries out a specific function within a eukaryotic cell. 2. Each plant cell has a cell wall and chloroplasts, while animal cells lack both structures. 3. A ribosome assembles proteins. 4. a. DNA is located in the nucleus of a eukaryotic cell. b. DNA is located in an area of the cytosol called the nucleoid in a prokaryotic cell. 5. Genes are made of DNA.

Climbing Up the Ladder: **Organelles and Cells**

In this chapter, we continue our climb up the hierarchical ladder of organization (**Figure 6.1**). We learned in Chapter 4 that all matter is made of chemical elements. The smallest piece of an element is an atom, and atoms can unite to form molecules. Large molecules, such as fat, DNA, starch, and proteins, are referred to as macromolecules. Atoms, molecules, and macromolecules provide the nutrients and building materials living organisms need to stay alive and healthy, and they participate in chemical reactions that do all of the work performed by those organisms.

This chapter will focus on the next two levels of organization: those involving cells. Macromolecules can unite to form complex structures called **organelles** that carry out specific functions within cells. Examples of organelles include the nucleus, a mitochondrion, or a chloroplast. **Cells** contain the combination of atoms, molecules, macromolecules, and organelles necessary to sustain life, so cells are the first level of organization that we consider to be alive.

Because most microorganisms are **unicellular**, consisting of just a single cell, the three levels above the cell—tissue, organ, and organ system—are not usually applicable. These are studied in anatomy and physiology of higher organisms. A **population** in microbiology consists of a localized group of a single species of microorganism. These populations combine with other biological populations (single-celled and **multicellular** organisms) to form a **community** (called a **biofilm** in microbiology) within a particular **ecosystem**.

✔ **QUICK CHECK**

Fill in the missing levels from the biological hierarchy of organization.
1. atom, molecule, _____, organelle
2. cell, _____, organ, _____, organism
3. _____, biofilm, ecosystem

Answers: 1. macromolecule 2. tissue; organ system 3. population

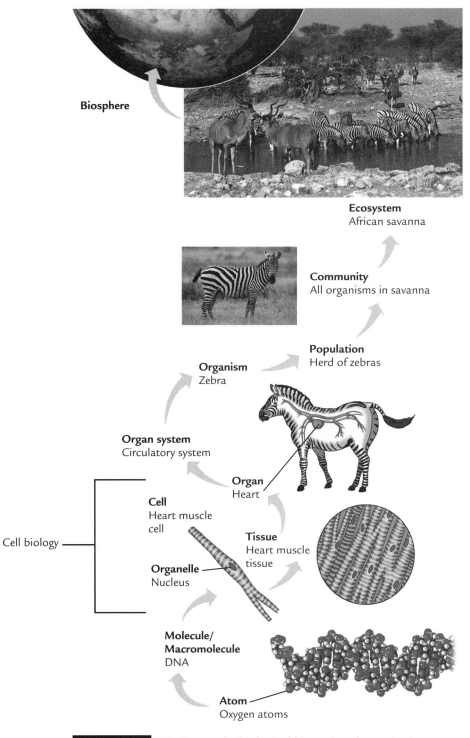

Biosphere

Ecosystem
African savanna

Community
All organisms in savanna

Population
Herd of zebras

Organism
Zebra

Organ system
Circulatory system

Organ
Heart

Cell
Heart muscle cell

Tissue
Heart muscle tissue

Cell biology

Organelle
Nucleus

**Molecule/
Macromolecule**
DNA

Atom
Oxygen atoms

FIGURE 6.1 Moving up the biological hierarchy of organization, we explore the organelle and cell in the discipline of cell biology.

As you learned in Chapter 5, all cells are either prokaryotic or eukaryotic. The prokaryotes are members of the domains Bacteria and Archaea; they are single celled. Eukaryotes make up the domain Eukarya, which includes all other organisms, from the single-celled protists and yeasts to multicellular fungi, animals, and plants. In this chapter, we will look more closely at the differences between the cells of prokaryotes and eukaryotes.

All bacteria are prokaryotic cells. All protists and fungi, animal, and plant cells are eukaryotic. ▪

WHY SHOULD I CARE?

We can use the differences between prokaryotic and eukaryotic cells as a tool in fighting disease. Antibiotics are drugs that we take to fight bacterial infections. Most antibiotics target a specific difference in structure or function between the bacterial cells and our cells. Thus, the antibiotic harms the bacterium while doing little harm to us.

Paper or Plastic? **The Cytoplasmic Membrane**

Prokaryotes and eukaryotes have some cell structures in common. One of these is the structure that encloses their cytoplasm: the **cytoplasmic membrane**, also called the **plasma membrane**. It is simultaneously a container that holds a cell together and a physical partition that separates the cell's inside world from the outside, making it easier to maintain homeostasis (relatively constant internal environment).

Imagine a zipped plastic bag filled with water as a model of a cell. That image works to illustrate the concept of the cytoplasmic membrane as a physical separation between the cell's inside and outside worlds. This separation makes it easier to maintain homeostasis inside the cell even if conditions outside the cell vary. But if the membrane was an actual bag and formed a complete barrier, nutrients could not enter cells, nor could wastes and products manufactured by cells leave.

Therefore, the cytoplasmic membrane must have unique characteristics that allow some materials to pass through while blocking others.

A cytoplasmic membrane is made primarily of **phospholipid** molecules. These molecules have phosphorus and oxygen at one end, forming what is called the **phosphate head**. This portion of the molecule is polar, or **hydrophilic**. Do you remember the meaning of "hydrophilic" from our discussion of polar molecules in Chapter 4? _____

_____

Attached to one side of the phosphate head are two longer molecules called **fatty acid tails**. These tails are the main reason that the molecule is a lipid, and you should recall from Chapter 4 that lipids are nonpolar, or **hydrophobic**. What does that mean? _____

_____

To envision how a phospholipid molecule is organized, think about a brass brad—an old-fashioned paper fastener you likely used in school or that may even be built into some of your folders now (**Figure 6.2**).

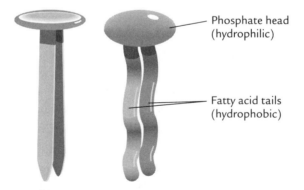

Phosphate head
(hydrophilic)

Fatty acid tails
(hydrophobic)

FIGURE 6.2   To understand the organization of a phospholipid molecule, think of a brass brad (diagram on left) that is used to hold together paper. There is a head with two long parts hanging off of it. The head represents the phosphate "head" of the molecule, which is hydrophilic. The prongs that hang down represent the fatty acid "tails," which are hydrophobic. The diagram on the right represents a model of a phospholipid.

TIME TO TRY

Let's try to figure out how the phospholipid molecules are arranged in a cytoplasmic membrane.

1. What happens when a lipid comes into contact with water? Fill a clear drinking glass or measuring cup about two-thirds full of water. Watching closely, pour 1/2 teaspoon of cooking oil into the water. Describe what happens when the oil first enters the water and where it settles. _____

   _____

2. Get a slice of bread and cover half of its surface with a thick layer of peanut butter, which has a high fat (lipid) content. Gently place one drop of water on the bare surface of the bread. What happens to the water? _____

   _____

   Gently place a drop of water on the peanut butter. What happens to the water? _____

   How do you think these observations relate to the organization of phospholipid molecules in a cytoplasmic membrane? _____

   _____

The phosphate heads of phospholipids interact fine with water, but the opposite ends of the molecules—the fatty acid tails—do not. As you saw in the first experiment, lipids in contact with water first form a ball, then settle into a single layer at the surface with one side in contact with the air, not water. In a cell, though, both surfaces are in contact with water. How can phospholipid molecules arrange themselves so the fatty acid tails do not contact water? _____

_____

Think about the peanut butter experiment. You should have seen the water readily enter the bare part of the bread but ball up on the peanut butter. How could you organize bread and peanut butter in such a way that there are two major surfaces, or sides, and water will not be repelled from either one? _____

_____

You guessed it—the basic peanut butter sandwich. Assume you made it with two slices of bread, each smeared with a thick layer of peanut butter and then sandwiched them together. You would have bread on both sides of the sandwich and peanut butter in the middle—two layers of it facing each other. This is the basic structure of the **phospholipid bilayer** (*bilayer* = two layers). The hydrophilic phosphate heads (bread) are arranged in two layers so that they face the water in the extracellular fluid and in the cytoplasm. The hydrophobic tails (peanut butter) are sandwiched in the middle, out of contact with the water. This is how cytoplasmic membranes are organized (**Figure 6.3**).

Cytoplasmic membranes don't contain only phospholipids—there are other molecules as well. Cytoplasmic membranes also contain pro-

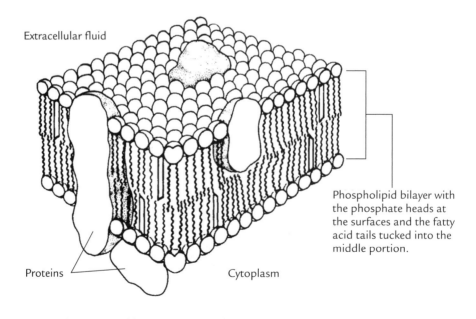

Extracellular fluid

Phospholipid bilayer with the phosphate heads at the surfaces and the fatty acid tails tucked into the middle portion.

Proteins

Cytoplasm

FIGURE 6.3   The phospholipid molecules are organized into a bilayer in which other molecules are embedded.

teins. Some of these proteins form channels that act like tunnels to allow certain molecules to pass through the membrane. Other proteins are carriers that pull various molecules through the membrane. The combination of phospholipid molecules and specialized protein channels and carriers determines what can and cannot pass through a cytoplasmic membrane. Because not everything can pass through, a cytoplasmic membrane is said to be **selectively permeable**—its chemical composition restricts the movement of some substances, so it is selective about what can pass through it.

In eukaryotes, cytoplasmic membranes contain sterol molecules that help stabilize the membrane. Eukaryotes also have carbohydrate molecules sticking out of the cytoplasmic membrane that are recognizable by other cells, such as when the immune system is trying to recognize foreign invaders to keep the body healthy.

All of these assorted molecules are positioned in what can be thought of as a sea of phospholipid molecules (see Figure 6.3). Imagine a child's pool filled with water, and floating in it are toy boats, inflatable toys, and a few children. The objects are free to move around in the water, and in fact they do. This is the nature of a cytoplasmic membrane. This model, known as the **fluid mosaic model**, reveals a cytoplasmic membrane that is dynamic, moving, changing, and fluid in nature.

A cytoplasmic membrane is composed of a phospholipid bilayer but also contains other molecules, and it is a very active structure. ∎

Before we move on to other cellular structures, let's summarize how molecules move in and out of the cytoplasmic membrane.

MOVEMENT THROUGH THE MEMBRANE

Molecules are constantly in motion. This motion is random unless there are other factors influencing movement in a certain direction. The relative amount of crowding of molecules (their concentration) affects how and where they move.

PICTURE THIS

Imagine holding a bottle of strong perfume. The perfume molecules are concentrated (crowded) inside the perfume bottle. They are bouncing off one another, but their movement is confined to the inside of the bottle. What happens when you open the bottle?

_____

_____

_____

The molecules are still bouncing off one another, but now they can bounce outside the bottle into the air of the room. A **concentration gradient** (see **Figure 6.4a**) exists between the high concentration of perfume molecules inside the bottle and the low concentration of perfume molecules outside the bottle. The molecules have moved spontaneously down their concentration gradient, from where they were most crowded to where they were least crowded (Figure 6.4a). This type of molecular movement is called **simple diffusion**. Diffusion continues until molecules are in equal concentration in both spaces (**Figure 6.4b**). At this point, the molecules are still moving but not in a particular direction. This state is called **equilibrium**.

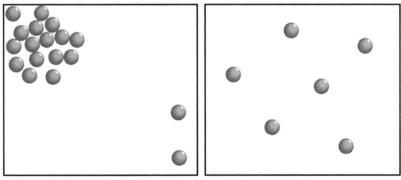

a) A concentration gradient   **b)** Equilibrium

FIGURE 6.4 **a)** A concentration gradient exists when there is a difference in the concentration, or spacing, around molecules in two different areas. **b)** Equilibrium exists when the molecules are spaced about evenly.

Oxygen is one molecule that moves into cells by simple diffusion through the lipid portion of the cytoplasmic membrane. It does so because it is a nonpolar molecule. However, larger polar molecules, such as glucose, cannot diffuse through the lipid portion of the membrane. Instead, they diffuse through protein carrier molecules that span the thickness of the membrane. This is a special type of diffusion called **facilitated diffusion**.

Diffusion is a passive process by which molecules spontaneously move from where they are in high concentration to where they are in low concentration. ▪

Osmosis is another type of special diffusion that is critical to cells' survival. It is the diffusion of water through a selectively permeable membrane, such as the cytoplasmic membrane. To understand this process, we need to define a solution and its components. A **solution** is a homogeneous mixture of a **solute** dissolved in a liquid **solvent**. A sugar solution, for example, consists of sugar (solute) dissolved in water (solvent). When a selectively permeable membrane separates two solutions that have a different concentration of solutes, there are two possible consequences. If the membrane is permeable to both solute and solvent, then both can diffuse along their concentration gradients. But, if the membrane is only permeable to the solvent (usually water), then only the water can diffuse. How do we determine which side of the membrane has the highest concentration of water? It's the side with the lowest concentration of solute.

TIME TO TRY

Look at **Figure 6.5**. In this illustration there is a higher concentration of solute molecules inside the cell than there is outside. Assume the solute molecules cannot pass through the cytoplasmic membrane.

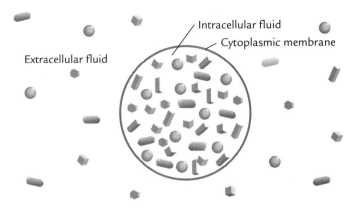

Extracellular fluid

Intracellular fluid

Cytoplasmic membrane

FIGURE 6.5 The cytoplasmic membrane separates the extracellular fluid from the intracellular fluid. In this figure, a situation is shown in which there are more solute molecules inside the cell than there are outside of it. Assume the shaded shapes are solute molecules and the remaining space is filled with water.

1. Which fluid contains the lowest concentration of solute molecules?

_____

2. Which fluid has the highest concentration of water? _____

_____

3. In which direction would water diffuse? _____

_____

Water diffuses from where it is in highest concentration (in this case, the extracellular fluid) to where it is in lowest concentration (inside the cell). Thus, water will diffuse into the cell.

WHY SHOULD I CARE?

One of the ways that we prevent microorganisms from spoiling food is to salt it (fish or other meats) or use lots of sugar (as in jams and jellies). This places the bacteria in an extracellular fluid that has more solutes (and less water) than their cytoplasm. These conditions would be the opposite of those seen in Figure 6.5.

Water diffuses out of the bacterial cells into the extracellular fluid, essentially dehydrating the bacterial cells. Without water, they cannot carry out metabolism or reproduce. Therefore, they will not spoil the food.

So far, we've only discussed movements that do not require any expenditure of energy by the cell. Sometimes the cell needs to move molecules against their concentration gradients. This requires that the cell use some energy, and the process is called **active transport**. Molecules and sometimes even cells can be transported into or out of a cell by surrounding them with a membrane. One of these processes is **phagocytosis** (*phago-* = "to eat"; *cyt-* = cell), which occurs when a cell engulfs another cell by surrounding it with extensions of its cytoplasmic membrane. Certain white blood cells use this process to "eat" the microorganism, digest it, and release the leftovers.

On the Assembly Line: **The Ribosome and Protein Synthesis**

The DNA within a cell has the instructions for all proteins that will be made in the cell, but the proteins are manufactured in a different location. A separate cellular structure called the **ribosome** is responsible for assembling the protein during a process called **protein synthesis**. You can think of a ribosome as being the assembly line on which the protein is built. The workers who build the proteins there are called **rRNA**, which stands for **ribosomal ribonucleic acid**.

The instructions or work assignments from the boss—the DNA—are carried to the ribosome by a special worker, a molecule called **messenger RNA (mRNA)**. Once the mRNA delivers the work order from the DNA to the ribosome, the specific proteins are made in the ribosome by linking together small molecules called amino acids. Amino acids are carried to the ribosomal assembly line by other workers called **transfer RNA (tRNA)**, which assemble them according to the instructions. So mRNA carries the instructions from the DNA to the ribosome, the tRNA brings the materials (amino acids), and the rRNA assembles them into proteins.

✔ QUICK CHECK

Match each "worker" with its job.

_____ 1. rRNA   a) Carries amino acids to the ribosome

_____ 2. tRNA   b) Assembles the amino acids into a protein

_____ 3. mRNA   c) Carries the instructions for protein assembly to the ribosome

Answers: 1. b 2. a 3. c

Both prokaryotic and eukaryotic cells have ribosomes. However, they are different in size and composition. Eukaryotic ribosomes are measured at a size called 80S (S = Svedberg unit, related to size), with 40S and 60S subunits. However, prokaryotic ribosomes are 70S (thus smaller), with 30S and 50S subunits.

WHY SHOULD I CARE?

Many antibiotics take advantage of this difference in ribosome structure and size. For instance, streptomycin changes the shape of the 30S ribosomal subunit in bacteria, preventing the correct reading of mRNA instructions for making a protein. Chloramphenicol prevents the 50S subunit from connecting (bonding) individual amino acids. These antibiotics affect the 70S ribosomes of bacteria, but they do not affect the cytoplasmic ribosomes of eukaryotes, which are 80S.

REALITY CHECK

Would streptomycin or chloramphenicol work on eukaryotic pathogens such as *Candida albicans*, a fungus that causes vaginal itching, or *Plasmodium*, a protozoan (protist) that causes malaria?

These antibiotics would not work because these eukaryotic pathogens have 80S ribosomes instead of the 70S ribosomes of bacteria.

What Department Are You With? **Cell Organelles of Eukaryotes**

Both prokaryotic and eukaryotic cells contain **cytoplasm** that fills the inside of the cell. The liquid part of the cytoplasm is called **cytosol**, a rather thick fluid in which many dissolved substances (such as nutrients and salts) and numerous cell structures are suspended.

Cells divide the labor among different specialized structures within the cytoplasm, each with a particular job. To understand this, think about a factory in which some product is manufactured. Inside this factory are many different departments, all involved in some way to make sure the factory is working properly and that the product is made and delivered. In a eukaryotic cell, many departments are the membranous organelles. Although each has its own particular task, all organelles are coordinated and work together. The main structures of eukaryotic cells are illustrated in **Figure 6.6**. Note that this figure shows several main differences between plant and animal cells. For example, plant cells have rigid cell walls and green chloroplasts, neither of which is found in animal cells.

THE NUCLEUS

The **nucleus** is the largest organelle and is found only in eukaryotic cells. It houses the DNA, which contains most genes. Each gene is essentially the instructions for how to make a specific protein. DNA is organized into threadlike strands called **chromosomes** when the cell reproduces.

The nucleus is enclosed in a **nuclear envelope** made of a double membrane, each of which is similar to the cytoplasmic membrane. This envelope is pierced periodically by holes called **nuclear pores** that allow larger molecules to pass through it. This is important because the instructions—the genes—never leave the nucleus. The boss sends work assignments out of the office (nucleus) in the form of molecules such as mRNA.

THE ENDOPLASMIC RETICULUM

The **endoplasmic reticulum (ER),** is an extensive network of membranous tubes and channels inside a eukaryotic cell. You can think of ER as being like a system of hallways in the factory, through which materials from one work station move to another. Because the ER connects

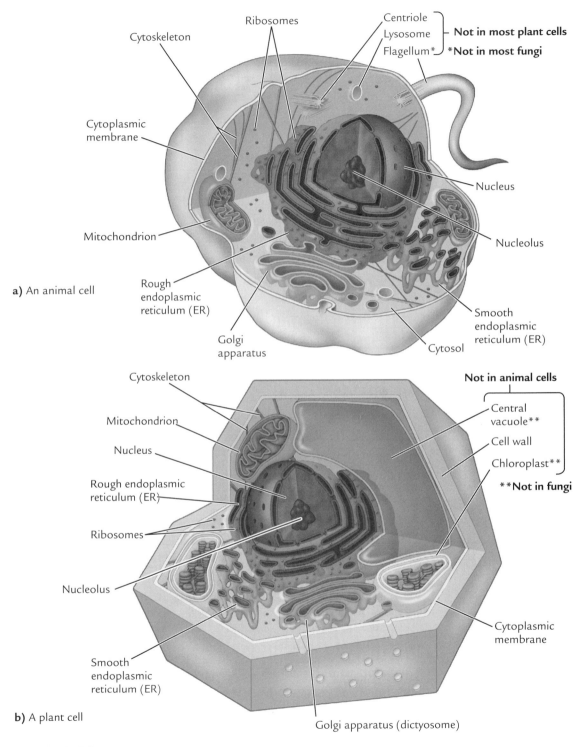

a) An animal cell

b) A plant cell

 FIGURE 6.6 Composite eukaryotic animal and plant cells showing the major organelles.

different parts of the cell, it also provides a communication network within the cell. Materials are brought into this system, moved around, changed, and turned into unrefined product here.

Some ER has ribosomes attached to the outside of the tubes, giving it a rough appearance; this is why it is called **rough ER**. The presence of ribosomes tells you that at least part of rough ER's job is protein synthesis. **Smooth ER** lacks ribosomes. Instead of making proteins, it is involved in making other materials (such as lipids), detoxifying potentially harmful substances, and transporting materials around the cell.

GOLGI APPARATUS

Under a microscope, the **Golgi apparatus** looks like a stack of flattened membranous sacs. This is the processing, packaging, and shipping department in cells. Products that have been made elsewhere in the cell are sent here to be modified and put into their final forms. They are slapped with a molecular "shipping label"—a chemical tag that determines where they will go. They are packaged in a membrane that pinches off of the Golgi apparatus and surrounds the product, forming a saclike structure called a **vesicle**. Finally the products are shipped to other parts of the cell, to the cytoplasmic membrane, or out to the great extracellular world beyond.

✔ **QUICK CHECK**

1. Where in the cell are proteins made? _____

2. What is the functional relationship between the nucleus, ribosome, rough endoplasmic reticulum, and Golgi apparatus?

_____

_____

_____

Answers: 1. ribosome 2. Nucleus contains instructions for making proteins, ribosome makes the proteins, rough ER transports proteins, and Golgi apparatus packages and ships them elsewhere.

THE MITOCHONDRION

If the factory is to do its work 24/7, it needs a good power source. The **mitochondrion** is a eukaryotic cell's powerhouse. It provides a constant supply of energy to drive the work being done throughout the cell. Energy in cells comes from food, where it is stored in the chemical bonds that hold the food's atoms together. Specialized chemical reactions in the mitochondria harness that chemical energy and put it in a molecule called **ATP**—*a*denosine *tri*phosphate. All cells use ATP directly for energy. Think of it as the rechargeable battery that powers cells. The more work a eukaryotic cell is doing, the more ATP it needs, and the more mitochondria it will have.

Mitochondria contain their own genetic information and can reproduce on their own. These elongated organelles (**Figure 6.7**) are enclosed by a double membrane. The inner membrane is highly folded into **cristae** (singular = *crista*). The folds increase the surface area of the membrane and thus increase the work space.

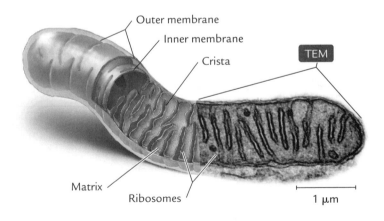

Outer membrane
Inner membrane
Crista
TEM
Matrix
Ribosomes
1 μm

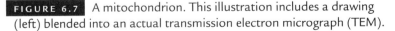

FIGURE 6.7 A mitochondrion. This illustration includes a drawing (left) blended into an actual transmission electron micrograph (TEM).

TIME TO TRY

To understand how the folds of the mitochondrial membrane increase the surface area, get a small plastic sandwich bag. This represents the outside membrane of the mitochondrion. Notice that it is smooth all the way around. If you put another smooth membrane inside of this bag, it would be about the same size. Now, take a larger plastic grocery bag and note how much larger it is. This will represent the inner folded membrane of a mitochondrion. Start folding it. Can you fold it enough to fit inside of the original bag? If you fold it enough, it should all fit inside. Now you can see how much more membrane is available if it is highly folded.

THE CHLOROPLAST

We just learned that living organisms use food as a chemical energy source. However, plants and other photosynthetic organisms also absorb light energy from the sun and convert it to the chemical energy of sugar and other organic molecules. Photosynthetic eukaryotic organisms are able to accomplish this because many of their cells, the ones that are green, contain unique organelles called **chloroplasts.**

Like mitochondria, chloroplasts contain their own DNA, can reproduce, and have a double membrane. They are partitioned into internal compartments (**Figure 6.8**). The **stroma** is the thick fluid within the

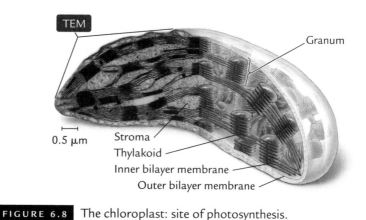

TEM

Granum

0.5 μm

Stroma
Thylakoid
Inner bilayer membrane
Outer bilayer membrane

FIGURE 6.8 The chloroplast: site of photosynthesis.

chloroplast. Suspended in that fluid, a network of membrane-enclosed tubes called **thylakoids** forms another compartment. These disks occur in stacks called **grana** (singular = *granum*), as shown in Figure 6.8. Molecules called **photopigments** located on the thylakoids actually trap the light energy.

LYSOSOMES

While the organelles are doing all this hard work in the cellular factories, they need some assistance in keeping their workplace tidy. **Lysosomes** are an animal cell's janitorial staff. A lysosome is a small membranous bag containing strong digestive enzymes. Its main job is to break down old, worn-out cell parts or foreign material that invades cells so they neither harm nor clutter the cell's interior. A lysosome recycles what it can and gets rid of the remaining garbage. Lysosomes are also involved in cell "suicide" or **apoptosis**. This process removes old or damaged cells.

THE CYTOSKELETON

Remember that all of this work is going on inside a cell, in a liquid environment. The "building" needs a frame to hold it up. We discussed the outer partition—the cytoplasmic membrane. But something is needed inside to hold the membrane out so the cell does not collapse on itself. A **cytoskeleton** is composed mostly of tiny tubes (**microtubules**) and filaments (**microfilaments**). These structures form a type of scaffolding that supports the cell and provides a place for various organelles to attach. Although the name sounds like this structure is made of bone, the cytoskeleton is actually made of proteins. You can think of them as being the struts and beams that hold up the cell.

CELL MOVEMENT: CENTRIOLES, CILIA, AND FLAGELLA

We have discussed how materials can move into and out of a cell, but other movements are associated with a cell, too. Paired cylindrical structures made of microtubules called **centrioles** direct the movement of the chromosomes when a cell reproduces. Microtubules also form part of two other structures involved in some eukaryotic cells' movement: cilia and flagella.

Under a microscope, **cilia** look like fringe on a cell. The movement of cilia is coordinated so that they tend to move in a wavelike manner. Cilia sweep materials over the outer surface of a cell, moving them away from the cell. For example, in the human respiratory tract, the cilia help clear debris so the particles don't get into and clog the air sacs where oxygen enters our blood. Some protozoa also have cilia that help them move. In contrast, a **flagellum** is a single, long, taillike extension of a cell. Human sperm cells and some protists have flagella to help them move from one location to another.

PICTURE THIS

Imagine you are at the "big game" and the crowd is tossing around a beach ball. As this is going on, the crowd starts a "wave," in which they stand and wave their arms overhead, then sit, group by group, all around the stadium. The wave approaches you just as the beach ball is heading your way. The wave passes you and so does the beach ball. It was carried away on the wave, and you see it now making its way around the stadium, riding the wave. This is how cilia move materials across the surface of the cell—they beat in a coordinated manner, sweeping materials along its length.

✔ QUICK CHECK

1. Flagella and cilia use tremendous amounts of energy for movement. Which organelle would you expect to see in large numbers near them? _____

2. If the cilia somehow miss some of the debris and it enters a cell, which organelle might destroy that debris? _____

Answers: 1. mitochondria 2. lysosome

We have reviewed the organelles of a typical eukaryotic cell. Now let's see how that compares with the prokaryotic cell—a much simpler type of cell.

The Tiny Factory: **How Prokaryotic Cells Differ**

Prokaryotes are much simpler in structure than eukaryotes. We can think of them as a tiny one-room factory. Look at **Figure 6.9,** which illustrates a prokaryotic cell. Note that there are no membranous organelles (such as mitochondria, chloroplasts, or Golgi apparatuses). In addition, because this factory is a single room, there are no hallways (endoplasmic reticulum). We have already learned that prokaryotes have ribosomes, the "assembly lines" that synthesize proteins, but the ribosomes are smaller than those found in the eukaryotic cytoplasm. Prokaryotic DNA (instructions for those proteins) is mostly contained

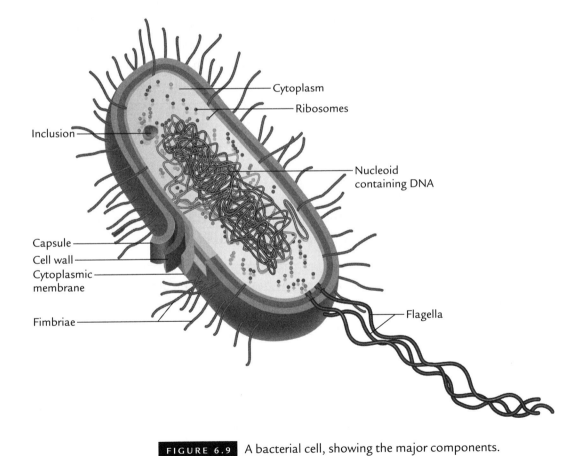

FIGURE 6.9   A bacterial cell, showing the major components.

within a single circular chromosome that is located in an area of the cytoplasm called the **nucleoid.** This area is not bound by membranes. Thus, the information for making proteins is in the same "room" as the "assembly line" in which the proteins are constructed.

OUTER COVERINGS

Now let's look at some other structures of the typical prokaryotic cell. We've already noted that all cells have a cytoplasmic membrane; the prokaryotic cytoplasmic membrane lacks sterols and outer carbohydrate "labels." In fact, the cytoplasmic membrane in most bacteria is not the outermost part of the cell. Most bacteria have a **cell wall** (Figure 6.9). The cell walls of organisms in the domain Bacteria are made primarily of a carbohydrate called peptidoglycan. Members of the domain Archaea lack peptidoglycan and have other molecules in their cell walls. Each cell wall gives shape to the cell but is quite porous. Bacteria can have one of three options with respect to the cell wall:

- **Gram-positive bacteria** have a thick wall, made of up to 40 interlocking layers of peptidoglycan (see **Figure 6.10a**).

- **Gram-negative bacteria** have a thin wall, made only of 1–2 layers of peptidoglycan. To the outside, there is an additional asymmetric bilayer called an outer membrane (see **Figure 6.10b**).

- **Mycoplasmas** are bacteria that lack cell walls. They have special molecules in their cytoplasmic membranes that make them more stable.

The "Gram" in the names of the first two groups comes from a procedure called the *Gram stain.* This staining technique results in purple cells if the cells are thick walled (Gram positive) and red cells if the bacteria are thin walled (Gram negative). You will undoubtedly practice the Gram-stain technique numerous times in your microbiology lab.

Bacterial cell walls usually have either teichoic acids or polysaccharides extending outward from the walls (see Figure 6.10). These act as identification "tags" called **antigens** that can be recognized by other bacteria or by our own defense cells.

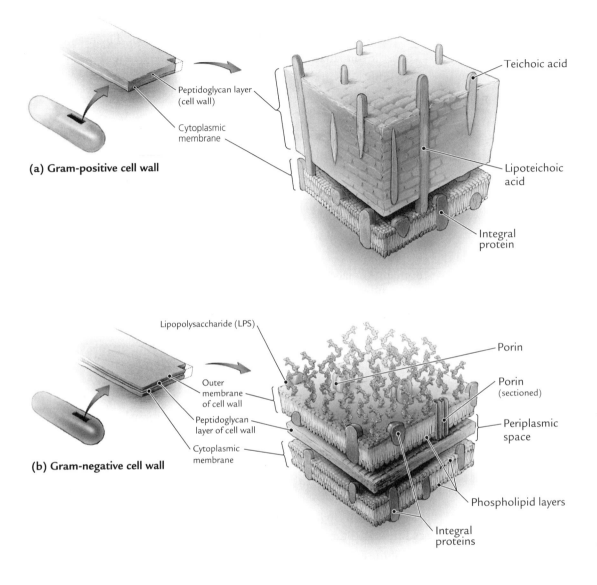

(a) Gram-positive cell wall

Peptidoglycan layer
(cell wall)

Cytoplasmic
membrane

Teichoic acid

Lipoteichoic
acid

Integral
protein

Lipopolysaccharide (LPS)

Outer
membrane
of cell wall

Peptidoglycan
layer of cell wall

Cytoplasmic
membrane

(b) Gram-negative cell wall

Porin

Porin
(sectioned)

Periplasmic
space

Phospholipid layers

Integral
proteins

FIGURE 6.10   Cell wall and cytoplasmic membrane of **a)** Gram-positive
and **b)** Gram-negative bacteria.

WHY SHOULD I CARE?

The antibiotic penicillin (and derivatives ampicillin and amoxicillin) disrupts the formation of the cell wall in a bacterium. This prevents bacteria from multiplying. Penicillin is useful because its mode of action does not affect human cells, considering our cells have no cell walls.

Some bacteria have a **glycocalyx** outside the cell wall. From its name, you can tell a glycocalyx is made of polysaccharides (*glyco* = "sugar"). In some bacteria, the glycocalyx takes the form of a fairly well-organized **capsule** (see Figure 6.9). Bacteria that possess capsules are often capable of evading detection by our defenses. For instance, *Streptococcus pneumoniae*, a bacterium that causes pneumonia, is unable to do so if it lacks a capsule. Our defense cells detect and kill the bacterium lacking a capsule before it can multiply.

In other bacteria, the glycocalyx is loosely attached and called a **slime layer**. Slime layers are also important to us. Bacteria can use them to form a **biofilm**, a complex community of bacteria that forms where liquid meets a solid surface, such as the surfaces of your teeth, the interior of medical tubing (catheters), or the insides of water pipes. Bacteria in biofilms communicate and collaborate with one another. Their glycocalyces also protect the biofilm from many methods of microbial control, including chemicals, radiation, and heat. Therefore, we are interested in learning more about these unique communities of bacteria.

APPENDAGES

Some bacteria have additional structures that attach to the cell wall and project outward from the cell. These are flagella, pili, and fimbriae. **Flagella** (singular *flagellum*) are the longest (see Figure 6.9) and are used to move the cell. They differ from eukaryotic flagella in their composition, size (about 1/10 the size of eukaryotic flagella), movement (they rotate like a propeller), and location (extracellular).

Fimbriae (see Figure 6.9) are the shortest of the appendages. They exist in large numbers along the entire surface of the cell, making it look as if it has bristles. Fimbriae help the bacterium to attach to a surface and even move along that surface. They are important in forming biofilms. **Pili** (singular *pilus*) are similar to fimbriae, but there are usually only 1–2 pili per cell, and they are longer. They help bacteria to attach to and transfer DNA to another bacterium.

✔ **QUICK CHECK**

Which appendage is the longest? _____

Which appendage(s) function in movement of bacterial cells?

_____

Which appendage helps to transfer DNA? _____

Answers: 1. flagellum 2. flagellum and fimbriae 3. pilus

INSIDE THE CELL

The interior of a bacterial cell (Figure 6.9) is fairly simple when compared with a eukaryotic cell (Figure 6.6). We've already discussed two of the bacterial cell's components: ribosomes and the nucleoid. Here are a couple of other structures that might be found inside bacterial cells: **inclusions** and **endospores**.

An inclusion is a type of deposit within the bacterial cell. These are often clusters of similar molecules that are waiting to be used by the cell—sort of like ingredients on your kitchen counter waiting to be put together into a meal. Inclusions can store sulfur, phosphates, and lipid polymers, or they can take the form of a gas vesicle containing nitrogen. Gas vesicles help some photosynthetic bacteria to float at the surface of the water where they can get more sunshine.

An endospore is a sort of "emergency survival pod" in which a bacterium stashes a copy of its DNA, a few important enzymes, and a little cytoplasm. An endospore is surrounded by thick layers of peptidoglycan and keratin (the protein that makes your fingernails hard). These endospores can survive extreme heat, chemicals, radiation, and dehydration that would normally kill the original cell. When conditions improve, endospores germinate (like seeds) and regenerate the original cell. Only a few species of bacteria produce endospores.

WHY SHOULD I CARE?

Because endospores are so resistant to most of the methods we use to control microorganisms, bacteria that form endospores are difficult to control. In food processing, we have to use pressure plus heat (e.g., a pressure cooker) in order to kill endospores. Boiling water alone won't kill them. If food is improperly preserved, endospores can then germinate, and the cells begin to multiply. These bacteria, such as *Clostridium botulinum*, can then secrete deadly toxins into the food.

✔ QUICK CHECK

What is the name of the disease caused by the organism *Clostridium botulinum*? (The specific epithet gives you a hint!)

Answer: botulism

Diseases such as tetanus (caused by *Clostridium tetani*) and botulism can be deadly because the bacteria produce toxins that affect the action of skeletal muscles (including those that help us breathe). Toxins of these two species have opposite effects. Tetanus toxin causes the muscles to contract but fail to relax. Botulism toxin prevents muscles from contracting.

You have seen that the structure of prokaryotic cells differs significantly from those of eukaryotes. In the final chapter of this book, we will use this basic information to learn more about prokaryotes and their importance.

Final Stretch!

Now that you have finished reading this chapter, it is time to stretch your brain a bit and check how much you learned.

WHAT DID YOU LEARN?

Try these exercises from memory first; then go back and check your answers, looking up any items that you want to review. Answers to these questions are at the end of the book.

PART A: ANSWER THE FOLLOWING QUESTIONS.

1. What is the basic difference between prokaryotic cells and eukaryotic cells?

_____

_____

2. What substance transmits information about characteristics from parents to offspring? _____

_____

_____

3. Describe the organization of the cytoplasmic membrane. _____

_____

_____

_____

4. Design a concept map for the following terms: protein, DNA, nucleus, ribosome, rough ER, Golgi apparatus, and export.

5. What is the difference between Gram-negative and Gram-positive cell walls?

_____

_____

_____

_____

_____

_____

6. How can we use structural differences between prokaryotic and eukaryotic cells to help us treat diseases? _____

_____

_____

_____

_____

_____

PART B: FILL IN THE MISSING INFORMATION IN THIS TABLE COMPARING PROKARYOTIC AND EUKARYOTIC CELLS. (The first row has been done as an example.)

Structure or Organelle	Prokaryote	Eukaryote
Nucleus	Not present; DNA located in region of cytoplasm called nucleoid	Present; bound by double membrane; encloses DNA into separate compartment
Cytoplasmic membrane	Present; lacks sterols and polysaccharides	1.
Ribosome	2.	80S; synthesizes proteins
3.	Present in a few species; "survival mode"; has thick wall that is resistant to heat, radiation, & chemicals	Absent
Endoplasmic reticulum (ER)	Absent; all cell components within cytoplasm ("one room")	4.
5.	Absent	Series of flattened membranes; used to label and package products for export
Mitochondrion	Absent; ATP synthesized on cytoplasmic membrane	6.
7.	Absent; photosynthesis occurs on cytoplasmic membrane or its extensions	Found in plants and algae; converts light energy to chemical energy of sugars
Lysosome	Absent	8.
Flagella	9.	Present in some cells; made of microtubules, extension of cell; moves in whiplike fashion

PART C: WRITE THE CHAPTER TERMS IN A NOTEBOOK AND DEFINE THEM IN YOUR OWN WORDS. GO BACK THROUGH THE CHAPTER TO CHECK YOUR MEANINGS, CORRECTING THEM AS NEEDED. LIST EXAMPLES WHEN APPROPRIATE.

Organelles

Cells

Unicellular

Population

Multicellular

Community

Biofilm

Ecosystem

Cytoplasmic (plasma)
    membrane

Phospholipid

Phosphate head

Hydrophilic

Fatty acid tails

Hydrophobic

Phospholipid bilayer

Selectively permeable

Fluid mosaic model

Concentration gradient

Simple diffusion

Equilibrium

Facilitated diffusion

Osmosis

Solution

Solute

Solvent

Active transport

Phagocytosis

Ribosome

Protein synthesis

Ribosomal RNA (rRNA)

Messenger RNA (mRNA)

Transfer RNA (tRNA)

Cytoplasm

Cytosol

Nucleus

Chromosomes

Nuclear envelope

Nuclear pores

Endoplasmic reticulum (ER)

Rough ER

Smooth ER

Golgi apparatus

Vesicle

Mitochondrion

Adenosine triphosphate
    (ATP)

Cristae

Chloroplasts

Stroma

Thylakoids

Grana

Photopigments

Lysosomes

Apoptosis

Cytoskeleton

Microtubules

Microfilaments

Centrioles

Cilia

Flagellum

Nucleoid

Cell wall

Gram-positive bacteria

Gram-negative bacteria

Outer membrane

Mycoplasmas

Antigens

Glycocalyces

Capsules

Slime layers

Flagella (bacterial)

Fimbriae

Pili

Inclusions

Endospores

7 Microbiology Basics

Tiny Organisms of Huge Importance

When you complete this chapter, you should be able to

- Name some organisms that are considered to be microorganisms.

- Explain the importance of microorganisms to our everyday lives.

- Describe the importance of and differences between viruses, viroids, and prions.

- Discuss why it is important to understand the growth requirements, metabolism, and genetics of bacteria.

- Explain how microbes cause disease and how we control them, both inside and outside of the body.

Your Starting Point

Answer the following questions to assess your microbiology knowledge.

1. What sorts of organisms would be called microorganisms?

   _____

   _____

2. Name some ways that we benefit from microorganisms.

   _____

   _____

3. What is a virus? _____

   _____

4. How do bacteria divide? _____

   _____

5. Name some ways that the human body defends against disease.

   _____

   _____

Answers: 1. Microorganisms are any organisms too small to be seen with the unaided eye, including bacteria, protozoa, some fungi, algae, and microscopic worms. 2. Microorganisms provide oxygen and other nutrients for the ecosystem, recycle wastes, convert harmful substances into less toxic forms, assist with food production, protect us from pathogens, and are used in industrial production of rubber, plastics, chemicals, and pharmaceuticals. 3. A virus is a tiny nonliving entity composed of protein and nucleic acid. 4. Bacteria divide using a process called binary fission. 5. Skin and mucous membrane barriers, normal microbiota, natural killer cells, phagocytes, inflammation, fever, antimicrobial substances, specialized lymphocytes, and antibodies

Now you're set to begin your study of microbiology. But guess what? You have already learned quite a bit about microbiology. You know that it is the study of microorganisms, or "tiny organisms." **Microorganisms** are those organisms too small to view with the unaided eye. You've also learned that most microorganisms are classified within two of the three domains: domain Bacteria and domain Archaea. Recall that both of these domains include single-celled (unicellular) organisms that are prokaryotic, meaning they have no true nucleus or membranous organelles. Bacteria and Archaea differ in a number of ways, including cell wall structure. Some microorganisms are members of the domain Eukarya, such as the protozoa and algae. Other eukaryotic microorganisms are found in the kingdom Fungi (yeasts, molds) and even kingdom Animalia (intestinal worms). **Figure 7.1** illustrates some of the varied forms of microorganisms.

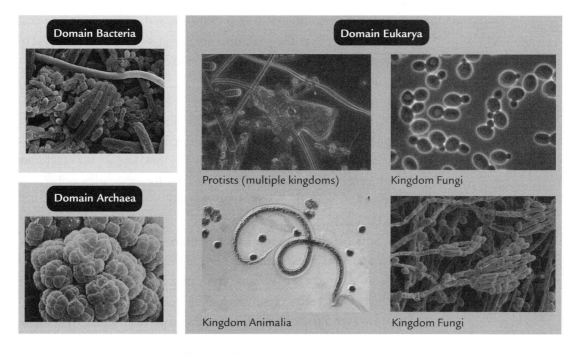

FIGURE 7.1 Types of microorganisms

✔ **QUICK CHECK**

What determines whether an organism is considered to be a "microorganism"? _____

_____

Answer: A microorganism is an organism that is too small to see with the unaided eye.

This final chapter will focus on some important topics in microbiology that will likely parallel the order of your microbiology course. In each section, you will see a preview of what you will learn in that unit, the importance of that topic, and pointers on how to learn it.

Good Things in Small Packages: **Importance of Microorganisms**

Microorganisms are the largest group of living organisms on the planet. We tend to first think of their importance as **pathogens**, organisms that cause disease. The pathogenic microorganisms tend to give the rest of the microbial world a bad reputation. However, most microorganisms are beneficial to us.

In fact, we probably couldn't exist on this earth without microorganisms. For example, certain bacteria called **cyanobacteria** are an important component of the plankton that live within the oceans and other bodies of water. They carry out photosynthesis and therefore are the producers in their ecosystem. In the process of photosynthesis, these cyanobacteria produce the oxygen that many other organisms need to survive.

Fungi and bacteria help to convert waste products into useful molecules. Some of these microorganisms, called recyclers, convert organic molecules (from dead plants and animals) into inorganic molecules that can be used by other organisms. Thus, they form the final "link" in the recycling of nutrients that prevents us all from being buried in garbage! There are even a few bacteria that can digest petroleum, thus helping to clean up oil spills.

REALITY CHECK

Can you think of some foods that are produced by microorganisms?

_____

_____

Bacteria and fungi are important in the food and beverage industries. You may already know that cheese is produced by molds (fungi), bread rises because of yeast (a fungus), and beverages such as wine and beer are produced by microorganisms. However, an abundance of other food products from many cultures are the result of some sort of fermentation by microorganisms: soy sauce, sauerkraut, kimchee, saki, sour cream, yogurt, vinegar, pickles, sausage, and even chocolate.

Industries also make use of microbial metabolism for the production of such things as synthetic rubber, polymers for plastics, and chemicals. The biotechnology industry uses both bacteria and viruses to genetically engineer and mass-produce pharmaceuticals such as insulin, growth factors, as well as other drugs and antibodies.

WHY SHOULD I CARE?

Antibacterial products have become popular in our society. People seem to have a fear of "germs." However, bacteria (and fungi) reside on our skin, as well as inside our upper respiratory, digestive, and reproductive tracts, and they help to protect us against disease. These microorganisms are termed our **normal microbiota**. In the large intestine, bacteria help to consume undigested food material and also produce vitamins that we use. On the skin and in the linings of many body systems, bacteria and fungi cover binding sites where pathogens might otherwise attach. Normal microbiota even secrete toxins that destroy invading bacteria. Finally, the presence of normal microbiota helps to keep our immune system vigilant.

A Microbe by Any Other Name: **Naming Microorganisms**

You'll be reading, writing, and talking about microorganisms in this course. Therefore, you will need to know how to properly name them. Let's begin with bacteria. The word "bacterium" is the singular form. You'd use it when talking about a single organism. "Bacteria" is the plural form, and it is used when referring to multiple organisms.

Now let's look a little further into the classification of microorganisms and their scientific names. Note that each kingdom within a domain is further divided into successively smaller groups (**Figure 7.2**). Eventually each microorganism has a two-part name, consisting of genus and specific epithet. For instance, the bacterium *Escherichia coli*, more commonly referred to as *E. coli*, is probably a name you've heard before. The *Escherichia* is the **genus** part of the name. The second word "*coli*" is the **specific epithet.** Together the two words describe a particular organism, a **species**. So *E. coli* is a species of bacterium. This type of naming system is called **binomial nomenclature** (*bi-* Latin for "two," *nom-* Latin for "name").

Rules for nomenclature specify that both words be spelled out completely when first used. After that, the genus name may be abbreviated with the first letter, hence *E. coli*. It is also required that a species name be italicized when typed and underlined when handwritten. This is part of what distinguishes it as a proper scientific name.

TIME TO TRY

You can often tell a lot about a microorganism by its scientific name. See if you can guess the significance of the following:

1. In which body organ would you find *E. coli*?

    _____

2. What color is the growth of *Micrococcus roseus*?

    _____

3. What disease do you think is the result of the following organisms?

    a. *Mycobacterium tuberculosis*

    b. *Streptococcus pneumoniae*

    c. *Neisseria meningitidis*

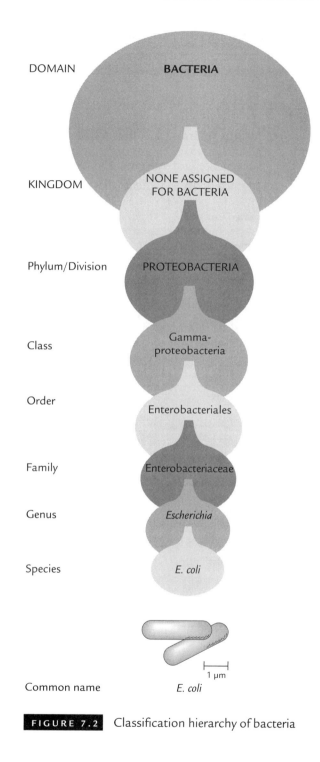

DOMAIN

BACTERIA

KINGDOM

NONE ASSIGNED
FOR BACTERIA

Phylum/Division

PROTEOBACTERIA

Class

Gamma-
proteobacteria

Order

Enterobacteriales

Family

Enterobacteriaceae

Genus

Escherichia

Species

E. coli

1 μm

Common name

E. coli

FIGURE 7.2 Classification hierarchy of bacteria

Sometimes the bacterial name tells us what the cells look like under the microscope. Bacterial cells usually have one of these two morphologies, or shapes (**Figure 7.3**):

■ **Coccus** (plural = *cocci*) is spherical

■ **Bacillus** (plural = *bacilli*) is rod shaped

Bacterial cells of a given species are arranged in a specific manner. We often use a prefix to indicate that arrangement:

■ *Mono-* (single)

■ *Diplo-* (in pairs)

■ *Strepto-* (in chains)

■ *Staphylo-* (in clusters)

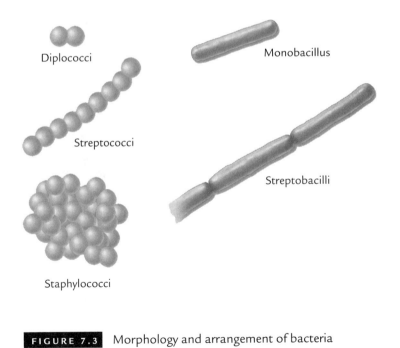

Diplococci

Monobacillus

Streptococci

Streptobacilli

Staphylococci

FIGURE 7.3   Morphology and arrangement of bacteria

TIME TO TRY

Draw the shape (and also arrangement, if you can tell) of the following genera of bacteria.

Streptococcus   *Staphylococcus*   *Thiobacillus*

Scientific names of bacteria may also tell us other things about that species. *Thiobacillus denitrificans*, which we now know takes the shape of a bacillus, is a great example. From Table 3.1, we know that the prefix "*thio-*" means sulfur. Maybe this bacillus uses sulfur as an energy source. The specific epithet *denitrificans* implies that the bacterium carries out a type of chemical reaction called denitrification, a form of nitrate reduction. From the scientific name alone, we deduce that it uses sulfur and reduces nitrate.

What Is It? **More Microbes**

In this chapter, we've concentrated on bacteria. We call them *microorganisms* because they have all of the features required for living organisms. However, microbiologists also study viruses, viroids, and prions. All three of these entities do not qualify as "living" because they are not made of cells, they lack the ability to reproduce on their own, and they must use their host's energy sources, building-block molecules, and metabolic machinery in order to replicate. So, we often use the term *microbes* to include viruses, viroids, prions, and bacteria.

A **virus** (see **Figure 7.4**) consists of a nucleic acid (either DNA or RNA) wrapped in a protein coat. Some viruses also pick up an envelope made of one of their host's membranes as they leave the host cell. Many diseases are caused by viruses, including flu, rabies, chicken pox, warts, genital herpes, and AIDS. You may also learn about viruses called **bacteriophages**

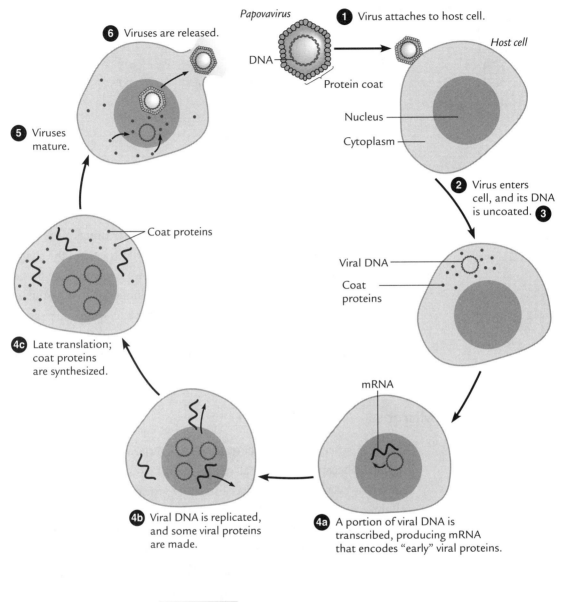

FIGURE 7.4 Typical virus life cycle

that have bacteria as their hosts. Figure 7.4 shows an example of a viral life cycle with the following components:

1. Attachment (a specific attraction between a molecule on a virus and a receptor molecule on a host cell).

2. Penetration (all or some of the virus enters the host cell)

3. Uncoating (protein coat is removed if entire virus enters host cell)

4. Biosynthesis (of viral nucleic acid and proteins)

5. Assembly (putting together the parts of the virus)

6. Release (virus bursts out of host cell or "buds" out within an envelope of the host membrane)

Viroids simply consist of nucleic acid, while **prions** only consist of protein. Viroids cause some plant diseases, while prions cause animal diseases, including bovine spongiform encephalopathy (mad cow disease) and Creutzfeldt-Jakob disease (CJD).

Viruses, viroids, and prions pose their own special problems when it comes to treating the diseases they cause. Considering they rely upon their host cells for replication, any medication is likely to have a significant detrimental effect on the host cells. For instance, the "cocktails" of several drugs that HIV-infected people must take several times a day have serious side effects.

✔ **QUICK CHECK**

Place a checkmark next to those features that are characteristic of viruses.

____ have nucleic acid   ____ can reproduce on their own

____ can make their own   ____ can break down sugar for
       proteins   energy

Answer: Viruses have nucleic acid but cannot do any of the other things listed.

Considering viruses, viroids, and prions are nonliving, the question is how to name and classify them. Taxonomists (scientists who specialize in classification) who study viruses are beginning to develop a nomenclature system similar to that for bacteria. However, it is not used universally, and you will notice a lot of irregularities in how viruses are named.

Tiny Factories: **How Bacteria Work**

In Chapter 6, we compared bacteria to small, one-room factories. All of the things necessary for a bacterial cell to function properly are enclosed within a single area—the cytoplasm. The instructions (DNA) for making proteins are in the same "room" as the assembly line (ribosomes). Part of your microbiology course will investigate how bacteria grow and convert energy to do their work.

GROWTH AND NUTRITION

You will find that, compared with cell division in eukaryotic cells, bacterial cell division is simple. The process is called **binary fission** and is shown in **Figure 7.5**. The bacterial cell replicates its DNA and then sep-

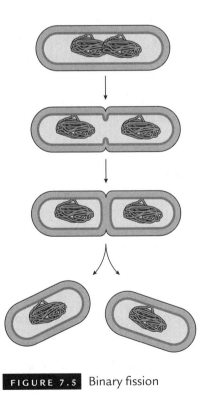

FIGURE 7.5 Binary fission

arates its cytoplasm by forming a new wall and membrane between the two daughter cells. Because the original cell has just one chromosome, this is a simple process.

Like all cells, bacteria have nutrition and growth requirements. Learning about these requires learning many terms. Sometimes the terms are grouped together, making for long words, such as chemo-lithoautotroph—a combination of chemotroph, lithotroph, and autotroph. Table 7.1 gives a preview of some of these terms. Before you look at the table, try this quick review of some important prefixes.

✔ **QUICK CHECK**

What is the meaning of each of these prefixes or suffixes? (Use Table 3.1 if you are not sure)

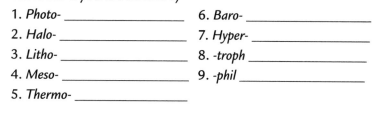

1. *Photo-* _____   6. *Baro-* _____
2. *Halo-* _____   7. *Hyper-* _____
3. *Litho-* _____   8. *-troph* _____
4. *Meso-* _____   9. *-phil* _____
5. *Thermo-* _____

Answers: 1. light 2. salt 3. rock 4. middle 5. temperature 6. pressure 7. over, above, excessive 8. food 9. loving

PICTURE THIS

Let's practice using some of these terms. Imagine a microorganism living in the deepest part of the ocean, near a thermal vent that is extremely hot. There are lots of organic molecules in dead organisms that have settled on the ocean bottom. There is no oxygen or light at this depth. What terms from Table 7.1 could be used to describe this organism's growth requirements?

_____

_____

TABLE 7.1

Growth Requirement	Associated Terms
Energy source	Phototroph (light) Chemotroph (chemicals)
Carbon source	Autotroph (CO_2) Heterotroph (organic carbon, such as sugars)
Electron source	Organotroph (organic molecules) Lithotroph (inorganic molecules)
Oxygen	Obligate aerobes (require oxygen) Microaerophiles (require oxygen in smaller amounts) Facultative anaerobes (prefer oxygen but can survive without it) Aerotolerant (don't use oxygen; can live with or without it) Obligate anaerobes (killed by oxygen)
Temperature	Psychrophiles (grow best below 15°C) Mesophiles (grow best at about body temperature, 37°C) Thermophiles (grow best at 60–70°C) Hyperthermophiles (grow best above 80°C)
pH	Acidophiles (grow best at acid pH) Neutrophiles (grow best at neutral pH) Alkalinophiles (grow best at alkaline pH)
Osmotic pressure	Osmotolerant (tolerate life in hypertonic solutions) Halophiles (prefer high salt environments)
Hydrostatic pressure	Barophiles (prefer high pressure environments)

How did you approach this question? Did you reread the scenario and look for clues? These are the clues that should have helped you answer the question:

- "Deep" implies hydrostatic pressure (from all the water above) is high → barophile

- "Extremely hot" → hyperthermophile

- "Organic material" → heterotroph, organotroph

- "No oxygen" → obligate anaerobe

- "No light" → chemotroph

So, you might ask, how do I prepare for such a question on a test? The best way is to know each term's meaning and then think of real-life examples that might fit that term. For example, where on the human body might you find halophiles? _____ (The skin is often salty because of excretion of salts through the pores during perspiration.)

METABOLISM

Now that you know a little about the requirements for growth of microorganisms, we can introduce some of the concepts of bacterial metabolism. It will help a lot if you've already had a biology course that discussed the metabolism of eukaryotes. However, the important thing is to focus on the "big picture" for each metabolic process or pathway you learn. Ask yourself the following helpful questions:

- What does the bacterium gain from this process?

- Where in the bacterial cell does this process take place?

- How does this process connect with other processes?

- How does this process in bacteria compare with that of eukaryotes?

- Do all bacteria do this, or is it a process that is unique to certain groups of bacteria?

As we learned in Chapter 4, metabolism can be divided into two parts: anabolism (building large molecules from smaller ones, using energy) and catabolism (breaking apart large molecules into smaller ones, releasing energy). In microbiology, you may study some or all of the following metabolic processes.

- **Glycolysis** begins with a 6-carbon sugar (such as glucose) and, through several chemical reactions, breaks it into two 3-carbon pyruvate molecules. This process yields adenosine triphosphate (ATP) which is an immediate energy source and an electron "ferry" called nicotinamide adenine dinucleotide (NADH).

- Glycolysis can be followed by the **transition step** and **Krebs (citric acid) cycle**. These processes begin with pyruvate molecules from glycolysis and yield ATP, CO_2, and electron "ferries" NADH and $FADH_2$.

- The **electron transport chain (ETC)** is where the electron ferries drop off their electrons and protons. The process ultimately yields ATP. The combination of glycolysis, transition step, Krebs cycle, and aerobic ETC is called **aerobic respiration** because oxygen is required. Some organisms conduct anaerobic respiration, in which oxygen is not used, and water is not formed.

- **Fermentation** may follow glycolysis and is an alternative to respiration. It does not require oxygen. Many different fermentation pathways use pyruvate from glycolysis and convert it to a variety of acids or alcohols. NADH (the electron ferry) drops off its electrons here and returns "empty" (in the form of NAD^+) to pick up more electrons from glycolysis.

- **Photosynthesis** is a process in which energy from sunlight is transferred to electron ferries that ultimately transfer it to form bonds between CO_2 molecules, producing organic molecules. Oxygen is often produced in this process. Plants and algae, of course, perform this process, but so do several types of bacteria, including cyanobacteria.

- Miscellaneous catabolic processes such as sulfur reduction or oxidation, nitrate reduction or oxidation, nitrogen fixation, and methane production are also possible for some bacteria. The list goes on and on!

Of course, these are just summaries. You will learn more details for each of these processes.

TIME TO TRY

When you begin this section of your course, try using a concept map to organize the topics into a big picture. Let's create a concept map that includes these terms: glycolysis, transition step/Krebs cycle, ETC, fermentation, CO_2, electron ferries, pyruvate, ATP.

To begin the map, write the terms for processes (e.g., glycolysis, transition step/Krebs cycle, ETC, fermentation) on 3" x 5" cards. Then think about which molecules are made or used in each process and make cards for them. Place the molecule cards near the card for the process that produces or uses each molecule. Move the cards around and talk about the processes and molecules (using the words "made" or "used").

The concept map in **Figure 7.6** illustrates the relationship of glycolysis, transition step/Krebs cycle, ETC, and fermentation. Notice that after glycolysis, a bacterium could continue to the transition step/Krebs cycle followed by ETC, it could go instead to fermentation.

GENETICS

Many of you will study the genetics of bacteria in your microbiology course. The basic diagram in **Figure 7.7** shows the flow of information from DNA. Recall in Chapter 6 we spoke of the eukaryotic cell as a fac-

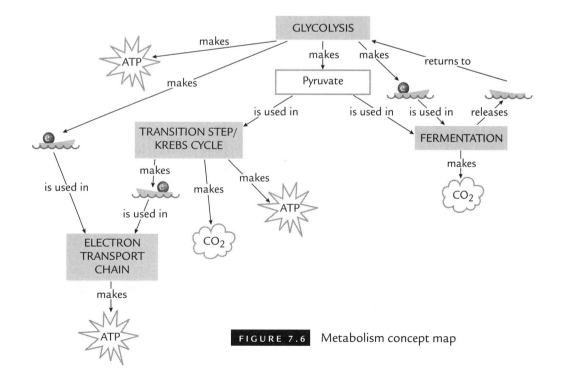

FIGURE 7.6   Metabolism concept map

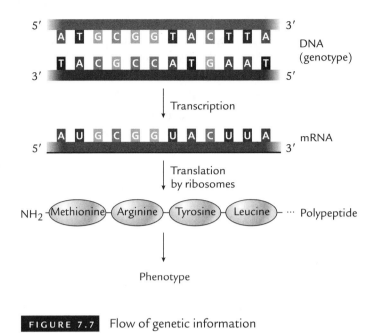

FIGURE 7.7 Flow of genetic information

tory, with DNA being the source of information (the "boss") for making the product. The information had to be carried from the boss's office (nucleus) out to the factory workers (ribosomes) where the proteins were made. However, prokaryotic cells have no nucleus, so we'll think of them as small, one-room factories. The boss (DNA) is right there in the same room with the rest of the workers. The boss still provides instructions in the form of mRNA. These instructions are easily found by the nearby ribosomes who use them to make proteins. In fact, these ribosomes are so close that they can begin to read the directions on the mRNA and begin making proteins before the entire instruction is completed! **Transcription** (creation of mRNA from DNA) and **translation** (creation of proteins by ribosomes from the mRNA instructions) happen almost simultaneously.

 Transcription (production of mRNA) and translation (production of proteins) both occur in the cytoplasm of bacteria. ▪

✔ **QUICK CHECK**

When bacteria make mRNA, this is called _____.

When bacterial ribosomes use the mRNA instructions to make a protein, this is called _____.

Answers: transcription; translation

Although bacteria don't carry out sexual reproduction, there are several methods by which new characteristics or traits can be acquired. One method is by **mutation**, which is a random error in DNA replication. Sometimes the frequency of these errors can be increased by outside factors, such as radiation or chemical agents called **mutagens**.

Bacteria can also gain new genes through several methods of **gene transfer** between two bacteria. There are three major methods of gene transfer shown and described in **Figure 7.8**:

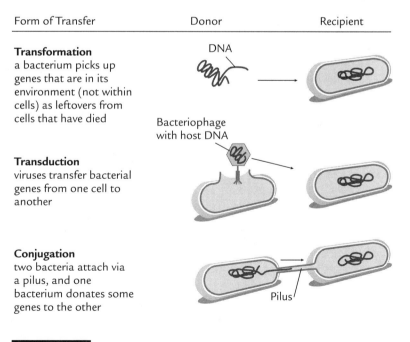

Form of Transfer	Donor	Recipient
Transformation a bacterium picks up genes that are in its environment (not within cells) as leftovers from cells that have died	DNA	
Transduction viruses transfer bacterial genes from one cell to another	Bacteriophage with host DNA	
Conjugation two bacteria attach via a pilus, and one bacterium donates some genes to the other	Pilus	

FIGURE 7.8   Methods of gene transfer

WHY SHOULD I CARE?

Recall that some bacteria can have genes that code for resistance to a particular antibiotic. These genes can, of course, be passed to the next generation when the cell divides. However, they can also be passed by gene transfer, sometimes even between different species of bacteria. This is one major reason why it is important not to use antibiotics for unnecessary purposes, such as taking antibiotics for a viral infection (antibiotics don't work on viruses) or "just in case" when you are traveling. Remember that antibiotics actually select for the resistant forms of bacteria, killing those that don't have resistance and allowing more space and nutrients for the resistant ones to multiply. You may have heard of **MRSA**, which stands for **m**ethicillin-**r**esistant *Staphylococcus **a**ureus*, a bacterial strain that can infect wounds and cause pneumonia. It is highly resistant to antibiotics and is difficult to treat.

PICTURE THIS

In this section, we've used four terms that begin with *"trans-."* One of your goals will be to develop strategies to distinguish similar terms when studying and taking exams. Here are a few suggestions:

- Transcription: to transcribe your notes, you copy them using the same language. When mRNA is made, essentially DNA is being copied in the same language ("nucleic acid language").

- Translation: a translator converts words from one language into another. When ribosomes perform translation, they begin with mRNA ("nucleic acid language") and translate it into a series of amino acids to make a protein ("amino acid language").

- Transformation: this literally means to change. When a bacterium picks up nucleic acid from its surroundings, it becomes changed.

■ Transduction: this also involves picking up nucleic acid from another organism. In this case, however, there is a virus that carries the genes from one cell to another. You might think of this as the virus "ab<u>duct</u>ing" (kidnapping) the genes and taking them to another location (ab<u>duction</u> . . . trans<u>duction</u>).

Gotcha! **Microbial Control**

Once you've learned about the different microorganisms, their growth requirements, metabolism, and genetics, then you have all the tools necessary to understand how we control microbial growth. This section of your course will focus on both physical and chemical agents of control, along with chemotherapeutic agents (such as antibiotics) used to treat disease. While this material might appear scary, you can understand how an agent is chosen by using common sense and the knowledge you've accumulated so far about the characteristics of microbes.

TIME TO TRY

To demonstrate how common sense plays a role in understanding the agents of control, see if you can answer these questions:

1. When you've had blood drawn, what types of microbial control were used? _____

    _____

2. If you were going to sterilize some plastic pipettes to use in an experiment, would you use a heat or gas sterilant? _____

    _____

3. If a patient has a type of pneumonia caused by a mycoplasmal bacterium, would it be better for the physician to prescribe penicillin (which destroys the cell wall) or chloramphenicol (which affects protein synthesis in prokaryotes)? _____

    _____

In the first example, you probably mentioned that the phlebotomist (the person who usually draws blood for medical tests) wore gloves and protective equipment such as goggles or a face mask and lab coat. Did the phlebotomist wipe your skin with an alcohol swab prior to inserting the needle? In addition, he or she used a disposable needle and syringe from a sterile package that he or she disposed of in a special container. Did the phlebotomist put an adhesive bandage on the injection site? Another method of microbial control! The phlebotomist wouldn't have wiped the injection site with bleach and wouldn't have reused a needle on another patient.

Hopefully you chose the gas sterilant for the plastic pipettes. Plastic might melt if placed in a high heat setting such as an autoclave.

The patient with pneumonia is infected with a bacterium that has no cell wall. Thus, penicillin would be useless in controlling the infection. See . . . this is not so hard! You'll be amazed at what you already know or are able to easily figure out.

Let's Settle Here: **Principles of Disease**

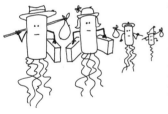

Your microbiology course will also include the ways in which microorganisms cause disease in their hosts. Think of them as pioneers or settlers, looking for a new home. Where would they choose to settle? Most likely a place that is similar to where they used to live. What do you know about what they need to grow? Do they like salt? Then they might settle on the skin. Do they like cooler temperatures? Then they might settle in the nose. Do they require anaerobic conditions (no oxygen)? Then they might settle deep within tissues where there is less oxygen available. Do they have flagella? Then they can move more easily from place to place. The pathogen will only attach to host cells if that cell has a particular receptor molecule that matches an attachment site on the pathogen. The pathogen may only enter the host cell if it has the means to do so. Some microorganisms that produce toxins don't even need to enter the body; instead, the toxin causes the illness (e.g., botulism).

Once these settlers have entered their new environment, they have to reproduce. Again, conditions you learned about in the growth and metabolism unit will be important for these microbes to reproduce. Do you think that a single bacterium or virus can cause disease? Not usually! Chances are that your defenses would destroy it. But if that bacterium can multiply before it's detected, then it's a bigger task for the defenses, and disease may result. Many pathogens have special features that allow them to escape detection by the defenses (e.g., capsules). This makes it more likely that they will be able to multiply and cause disease.

Disease itself can be studied in terms of the impact it makes on a population. This is called **epidemiology**. Epidemiology gets its name from the term *epidemic*, which means a sudden, unusually large number of disease cases within a particular time period. **Epidemiologists** study the spread of disease, how it is transmitted, and factors that influence that transmission before they develop methods to control that disease within the population.

In the epidemiology unit of microbiology, you will encounter a number of terms that describe diseases. One great way to learn these is to make flash cards with the term on one side and the description on the other. Another method that's fun would be to write a fictitious news story or fairy tale about a disease, using as many terms as possible in the story.

A significant part of your course will focus on specific body systems and the diseases associated with them. You may know many of these diseases already.

REALITY CHECK

What do you know about the effect of each of these diseases on the body?

1. Streptococcal pharyngitis ("strep throat") _____

2. Conjunctivitis ("pinkeye") _____

3. Viral gastroenteritis _____

You probably know that strep throat is an extremely sore throat, often so severe the patient cannot drink or eat comfortably. You might even know that this is caused by bacteria that are spherical and arranged in chains.

Note also that all three of the diseases mentioned end in "*-itis.*" This suffix means "inflammation" and tells you a little more about the disease. Conjunctivitis is inflammation of the conjunctiva of the eye. It can be very painful and is quite contagious.

Finally, gastroenteritis can be broken down into its root components: "*gastro-*" means stomach, "*entero-*" means intestines, and "*-itis*" means inflammation. Thus you might guess that viral gastroenteritis involves inflammation in the digestive tract, leading to vomiting and diarrhea. This is often heard of in conjunction with epidemics in nursing homes or on cruise ships. See how learning word roots can help you to speed up your learning of concepts?

TIME TO TRY

When studying diseases, it's good to make flash cards with the disease name on one side and the important features your instructor wants you to learn on the other. You can then practice these during little breaks at work or while you're sitting (not driving) in traffic. It is also helpful to group diseases by common agents, common mode of transmission, or common signs and symptoms. Then look for special ways to distinguish those similar groups.

The Defenders: **Human Defense Against Disease**

Your course might include a study of how the human body defends against disease. This defense process can be divided into two major categories: **innate defenses** and **adaptive defenses**. To discuss these defenses, let's use the analogy of a football game. The pathogens that invade the body are the "offense." The "defenses" are the parts of the body that try to keep the pathogens from going any farther into the body.

Figure 7.9 shows an overview of the defenses of the human body. The innate defenses are the "front line" of your body's defenses. Innate defenses (Figure 7.9) are present and acting most of the time. They are usually the first to encounter the invaders and recognize them as just general invaders (e.g., an RNA virus, a bacterium, or a Gram-negative

Innate (Nonspecific) Immunity		Adaptive (Specific) Immunity
First line of defense	**Second line of defense**	**Third line of defense**
· Intact skin · Mucous membranes and their secretions · Normal microbiota	· Natural killer cells and phagocytic white blood cells · Inflammation · Fever · Antimicrobial substances	· Specialized lymphocytes: T cells and B cells · Antibodies

FIGURE 7.9   Summary of human defenses

bacterium), not as a specific species. Innate defenses include the first-line defenses such as your skin, mucous membranes, and normal microbiota. If a pathogen escapes the first line, then the second-line defenses are ready to capture or kill it. These second line, innate defenses include white blood cells which act as **phagocytes** (cells that eat invaders), certain molecules such as stomach acid or interferon, and processes such as inflammation and fever. They affect a broad spectrum of pathogens and can act quickly.

Adaptive defenses are the defenses that act on specific pathogens. This third line of defenses must be activated before they can work. These defenses include certain white blood cells called **lymphocytes**, along with molecules called **antibodies** that are produced by some of these lymphocytes. Each lymphocyte is programmed in advance to recognize a particular chemical sequence (**antigen**) on the outside of the invader. Once that invader is recognized and the lymphocyte is activated (by a phagocyte), then the lymphocytes divide to produce cells that have a specialized job. Some of them produce antibodies (also programmed to recognize that specific invader), and others are capable of causing the invader to destroy itself. Adaptive defenses also have memory, meaning that some of these lymphocytes remain after the infection is over to carry the "memory" of that invader for the future. Should that specific invader return later (months, years, or even decades), there is still the ability to recognize it and respond quicker than the first time. This is why you usually don't get sick twice with the same disease, and it is the basis of what we call "**immunity**."

You will also learn in this section about vaccines, which give the recipient defenses without having the disease. You'll learn about some of the ways that the defenses can malfunction to create what we call "defense disorders." Such disorders include arthritis and allergies.

WHY SHOULD I CARE?

Medicine is all about disease. In order to be successful in your health care career, you need to understand the methods that the patient relies on to fight disease and the factors that affect their function. This will help you to better understand how to diagnose and treat each patient.

The End Is Your Beginning

Now you're ready to start your journey through the fascinating world of microbiology. Enjoy!

Final Stretch!

Now that you have finished reading this chapter, it is time to stretch your brain a bit and check how much you learned.

WHAT DID YOU LEARN?

Try these exercises from memory first; then go back and check your answers, looking up any items that you want to review. Answers to these questions are at the end of the book.

PART A: ANSWER THE FOLLOWING QUESTIONS.

1. Name some ways in which microorganisms are beneficial. _____
_____
_____
_____

2. Why isn't a virus considered a living organism? _____
_____
_____
_____

3. Which term describes an organism that uses sunlight for energy?

   a. Chemotroph   c. Heterotroph
   b. Phototroph   d. Halophile

4. Put these metabolic processes in order, with #1 being the first to occur within a bacterial cell.

   _____ Electron transport chain

   _____ Glycolysis

   _____ Transition step/Krebs cycle

5. Why can translation begin before transcription is finished in a bacterial cell but not in a eukaryotic cell? _____
_____
_____
_____
_____

6. Name the three methods of gene transfer and describe how each is unique.

_____

_____

_____

_____

_____

7. What is the difference between innate and adaptive defenses, and which one acts first?

_____

_____

_____

_____

_____

PART B: FOR EACH OF THE FOLLOWING ITEMS, MATCH THE PROCESS WITH ITS DESCRIPTION.

1. Transcription _____

2. Transformation _____

3. Glycolysis _____

4. Conjugation _____

5. Transduction _____

6. Binary fission _____

7. Fermentation _____

8. Photosynthesis _____

a) Producing of acids and alcohols from pyruvate

b) Transferring DNA by a virus

c) Dividing a bacterial cell

d) Converting light energy to chemical energy

e) Transferring DNA through a pilus connecting bacterial cells

f) Breaking a 6-carbon sugar into two 3-carbon pyruvate molecules

g) Copying information from DNA to a mRNA molecule

h) Taking up DNA from the environment

i) Using information in mRNA to make proteins

PART C: WRITE THE CHAPTER TERMS IN A NOTEBOOK AND DEFINE THEM IN YOUR OWN WORDS. GO BACK THROUGH THE CHAPTER TO CHECK YOUR MEANINGS, CORRECTING THEM AS NEEDED. LIST EXAMPLES WHEN APPROPRIATE.

Microorganisms
Pathogens
Cyanobacteria
Normal microbiota
Genus
Specific epithet
Species
Binomial nomenclature
Coccus
Bacillus
Microbes
Virus
Bacteriophages
Viroids
Prions

Binary fission
Glycolysis
Transition step and Krebs
    (citric acid) cycle
Electron transport chain
    (ETC)
Aerobic respiration
Fermentation
Photosynthesis
Transcription
Translation
Mutation
Mutagens
Gene transfer
Transformation

Transduction
Conjugation
MRSA
Epidemiology
Epidemic
Epidemiologists
Innate defenses
Adaptive defenses
Phagocytes
Lymphocytes
Antibodies
Antigen
Immunity

Answer Key

CHAPTER 1
Answers will vary from student to student.

CHAPTER 2

Part A
1. −32
2. 20,000
3. 3/4
4. 1/4
5. 1/4
6. 0.2
7. 1,000,000 or 10^6 per liter
8. 400 cm
9. 30.48 cm
10. 63 patients

Part B
1. 14
2. 4
3. 0.3; 30%
4. 15%
5. volume of a liquid
6. gram; meter (technically kilometer, but in practical use it is the meter); liter
7. There are four parts water to one part sugar, or four times as much water as sugar.
8. 100°C
9. volume
10. 1000

CHAPTER 3

Part A
1. e
2. f
3. i
4. j
5. g
6. h
7. b
8. d
9. c
10. a

Part B
1. d
2. b
3. glycolysis; exocytosis; microaerophile
4. The process of dividing the cytoplasm of a cell; "colored body"—condensed genetic material that can be stained and seen under a microscope; the study of tissues to diagnose a disease.
5. pharynges; mitochondria; flagella

CHAPTER 4

Part A

1. 19
2. 126.9 u
3. 8; 8
4. 7
5. 2; 8

Part B

1. protons and neutrons
2. electrons
3. The calcium atom lost two electrons, giving it a +2 charge.
4. In ionic bonds, the electrons physically move from one atom to another so that the atoms involved lose or gain electrons, and the opposite charges of the resulting ions draw the atoms together. In covalent bonding, the atoms share electrons.
5. $C_6H_{12}O_6$ and CH_2OH are organic molecules.
6. synthetic (anabolic)

CHAPTER 5

Part A

1. made of cells; maintain homeostasis; reproduce; interact with environment; convert energy; adapt; and evolve
2. Homeostasis is the maintenance of a constant internal environment. It is important to maintain homeostasis because cells perform optimally in a balanced internal environment.
3. DNA transmits information about characteristics from parents to offspring.
4. Most living things ultimately get their energy from the sun.
5. When a population of insects encounters pesticides in its environment, some insects will survive the exposure and live to produce offspring. These offspring will have the pesticide-resistant characteristics of their parents. Over time, the population of insects will contain a larger percentage of the pesticide-resistant individuals. This is an example of natural selection in action.
6. The three domains are Archaea, Bacteria, and Eukarya, and the three kingdoms are Animalia, Plantae, and Fungi. The Protista used to be a single kingdom, but now are divided into several kingdoms.
7. observation; question; hypothesis; testing

Part B

1. b
2. c
3. d
4. e
5. d
6. e
7. a

CHAPTER 6

Part A

1. Prokaryotes have no nucleus and lack internal membranes. Eukaryotes have both.
2. DNA
3. The cytoplasmic membrane is a phospholipid bilayer with protein channels periodically passing through it and other molecules "floating" in it.
4. Concept maps will vary from student to student, but these relationships should be included:

   - DNA contains the instructions for how to build proteins and is located in the nucleus.

   - Rough endoplasmic reticulum (ER) gets its appearance from the presence of ribosomes.

   - Proteins are made at ribosomes, so rough ER makes proteins.

   - Proteins from rough ER move to the Golgi apparatus for processing and packaging, then are exported.

5. Gram-positive cell walls are thick, while Gram-negative cell walls are thin and contain an outer membrane.
6. By targeting structures in prokaryotes that are different or unique, we can avoid effects on our own cells.

Part B

1. Present; contains sterols and polysaccharides
2. 70S; synthesizes proteins
3. Endospore
4. Present; transports materials
5. Golgi apparatus
6. Present; synthesizes ATP
7. Chloroplast
8. Present in animal cells; destroys unwanted materials
9. Present in some; made of flagellin; extracellular; rotational movement

CHAPTER 7

Part A

1. produce carbon compounds and oxygen; recycle waste into useful molecules, food, and beverages; contribute to industry and biotechnology; exist as normal microbiota
2. Viruses lack cellular structure and cannot reproduce or convert energy without using their host's machinery.
3. b
4. ETC = 3, glycolysis = 1, Transition step/Krebs cycle = 2
5. Eukaryotes have a nuclear envelope that separates the location for transcription (inside the nucleus) from the location for translation (ribosomes in cytoplasm). So the mRNA must be completed before it leaves the nucleus to participate in protein synthesis.
6. Transformation picks up DNA from the environment, while transduction uses viruses to transport DNA between bacterial cells, and conjugation uses a pilus to transport DNA between cells.
7. Innate defenses act first and recognize general features of invaders. Adaptive defenses act later and are very specific in what they recognize.

Part B

1. g
2. h
3. f
4. e
5. b
6. c
7. a
8. d

Photo and Illustration Credits

Unless noted, all chapter opener art and cartoon spot art was created by Kevin Opstedal.

CHAPTER 1

Table 1.2: All photos from Photodisc.

Figures 1.1, 1.2, 1.3: Seventeenth Street Studios.

Figure 1.4: Tom Steward/CORBIS.

CHAPTER 2

Figures 2.1, 2.2, 2.3, 2.4: Seventeenth Street Studios.

Figure 2.5: Lammert, *Techniques in Microbiology*, © Prentice Hall Publishing, Pearson Education, 2007.

Figure 2.6: Richard Megna/ Fundamental Photographs, NYC.

Figure 2.7: Adapted from Tortora/Funke/Case, *Microbiology: An Introduction, 9/e*, F21.13. © Pearson Benjamin Cummings, 2007.

Figure 2.8: Adapted from Tortora/ Funke/Case, *Microbiology: An Introduction, 9e*, F22.03. © Pearson Benjamin Cummings, 2007.

Figure 2.9: Adapted from Tortora/ Funke/Case, *Microbiology: An Introduction, 9e*, Table 14.5 (p.437). © Pearson Benjamin Cummings, 2007.

CHAPTER 3

Figure 3.1: Seventeenth Street Studios.

Figure 3.2: hip implant: National Institutes of Health by Seventeenth Street Studios.

CHAPTER 4

Figure 4.1: PhotoDisc/Getty Images; James Gritz; PhotoDisc/Getty Images/Adapted from Campbell, Reece, and Simon, *Essential Biology*, 3e, F1.3 © Benjamin Cummings, 2007.

Figures 4.2, 4.3, 4.4, Time to Try, 4.5, 4.6, 4.7, 4.8, 4.9, 4.10: Seventeenth Street Studios.

CHAPTER 5

Time to Try: Seventeenth Street Studios

Figure 5.1: Adapted from Campbell, Reece, and Simon, *Essential Biology*, 3e, F13.2a, © Benjamin Cummings, 2007.

Figure 5.2: Edward S. Ross, California Academy of Sciences/ Adapted from Campbell, Reece, and Simon, *Essential Biology*, 3e, F13.2c, © Benjamin Cummings, 2007.

Figure 5.3: Seventeenth Street Studios.

Figure 5.4a: Oliver Meckes/Nicole Ottawa/Photo Researchers/ Adapted from Campbell, Reece, and Simon, *Essential Biology*, 3e, F1.9, © Benjamin Cummings, 2007.

Figure 5.4b: Ralph Robinson/Visuals Unlimited/Adapted from Campbell, Reece, and Simon, *Essential Biology*, 3e, F1.9, © Benjamin Cummings, 2007.

Figure 5.4c: CORBIS/Adapted from Campbell, Reece, and Simon, *Essential Biology*, 3e, F1.9, © Benjamin Cummings, 2007.

Figure 5.4d: Digital Vision/Getty Images/Adapted from Campbell, Reece, and Simon, *Essential Biology*, 3e, F1.9, © Benjamin Cummings, 2007.

Figure 5.4e: CORBIS/Adapted from Campbell, Reece, and Simon, *Essential Biology*, 3e, F1.9, © Benjamin Cummings, 2007.

Figure 5.4f : D. P. Wilson/Photo Researchers/Adapted from Campbell, Reece, and Simon, *Essential Biology*, 3e, F1.9, © Benjamin Cummings, 2007.

Figure 5.5: Precision Graphics. Bauman, *Microbiology with Diseases by Body System, 2e*, F1.13. © Pearson Benjamin Cummings, 2009.

CHAPTER 6

Figure 6.1: PhotoDisc/Getty Images; James Gritz; PhotoDisc/Getty Images/Adapted from Campbell, Reece, and Simon, *Essential Biology*, 3e, F1.3, © Benjamin Cummings, 2007.

Figure 6.2, 6.3, 6.4, 6.5: Seventeenth Street Studios.

Figure 6.6: Adapted from Campbell, Reece, and Simon, *Essential Biology*, 3e, F4.6, © Benjamin Cummings, 2007.

Figure 6.7: Kenneth Probst. Bauman, *Microbiology with Diseases by Body System, 2e*, F3.36. © Pearson Benjamin Cummings, 2009.

Figure 6.8: Kenneth Probst. Bauman, *Microbiology with Diseases by Body System, 2e*, F3.37. © Pearson Benjamin Cummings, 2009.

Figure 6.9: Precision Graphics. Tortora/Funke/Case, *Microbiology: An Introduction,* 9e, F4.6a. © Pearson Benjamin Cummings, 2007.

Figure 6.10: Kenneth Probst. Adapted from Bauman, *Microbiology with Diseases by Body System, 2e,* F3.14. © Pearson Benjamin Cummings, 2009.

CHAPTER 7

7.1a: Steve Geschmeissner/Photo Researchers.

7.1b: Ralph Robinson/Visuals Unlimited.

7.1c: D. P. Wilson/Photo Researchers.

7.1d: Sinclair Stammers/Photo Researchers.

7.1e: L. Brent Selinger, Department of Biological Sciences, University of Lethbridge, Alberta, Canada; Pearson Education/Pearson Science.

7.1f: David Scharf/Peter Arnold.

Figure 7.2: Adapted from Tortora/Funke/Case, *Microbiology: An Introduction,* 9e, F10.5. © Pearson Benjamin Cummings, 2007.

Figure 7.3: Precision Graphics. Bauman, *Microbiology with Diseases by Body System, 2e,* F11.6, F11.7. © Pearson Benjamin Cummings, 2009.

Figure 7.4: Adapted from Tortora/Funke/Case, *Microbiology: An Introduction,* 9e, F13.15. © Pearson Benjamin Cummings, 2007.

Figure 7.5: Lee D. Simon/ Photo Researchers. Tortora/Funke/Case, *Microbiology: An Introduction,* 9e, F6.11a. © Pearson Benjamin Cummings, 2007.

Figure 7.6: Seventeenth Street Studios.

Figure 7.7: Precision Graphics.

Figure 7.8: Seventeenth Street Studios.

Figure 7.9: Adapted from N. Campbell, *Biology,* 4e, F39.1. © Benjamin Cummings, 1996.

Index